Building Android Apps in Python Using Kivy with Android Studio

With Pyjnius, Plyer, and Buildozer

Ahmed Fawzy Mohamed Gad

Apress®

Building Android Apps in Python Using Kivy with Android Studio

Ahmed Fawzy Mohamed Gad
Faculty of Computers & Information, Menoufia University, Shibin El Kom, Egypt

ISBN-13 (pbk): 978-1-4842-5030-3   ISBN-13 (electronic): 978-1-4842-5031-0
https://doi.org/10.1007/978-1-4842-5031-0

Managing Director, Apress Media LLC: Welmoed Spahr
Acquisitions Editor: Celestin Suresh John
Development Editor: Rita Fernando
Coordinating Editor: Divya Modi

Cover designed by eStudioCalamar

Cover image designed by Pixabay

Distributed to the book trade worldwide by Springer Science+Business Media New York, 233 Spring Street, 6th Floor, New York, NY 10013. Phone 1-800-SPRINGER, fax (201) 348-4505, e-mail orders-ny@springer-sbm.com, or visit www.springeronline.com. Apress Media, LLC is a California LLC and the sole member (owner) is Springer Science + Business Media Finance Inc (SSBM Finance Inc). SSBM Finance Inc is a **Delaware** corporation.

For information on translations, please e-mail rights@apress.com, or visit http://www.apress.com/rights-permissions.

Apress titles may be purchased in bulk for academic, corporate, or promotional use. eBook versions and licenses are also available for most titles. For more information, reference our Print and eBook Bulk Sales web page at http://www.apress.com/bulk-sales.

Any source code or other supplementary material referenced by the author in this book is available to readers on GitHub via the book's product page, located at www.apress.com/978-1-4842-5030-3. For more detailed information, please visit http://www.apress.com/source-code.

Printed on acid-free paper

*To this person who opens his mouth after
hearing that I wrote a book. Thanks for not believing in me.*

Table of Contents

About the Author

Ahmed Fawzy Mohamed Gad received his Master's of Science degree in 2018 and his Bachelor's of Science in Information Technology with honors in 2015. Ahmed is a researcher who is interested in deep learning, machine learning, computer vision, and Python. He is a machine learning consultant helping others with projects. Ahmed contributes written tutorials and articles to a number of blogs, including KDnuggets, Heartbeat, and Towards Data Science.

Ahmed authored two books about artificial neural networks and deep learning—*TensorFlow: A Guide to Build Artificial Neural Networks Using Python* (Labmert 2017) and *Practical Computer Vision Applications Using Deep Learning with CNNs* (Apress, 2018). He is always looking to learn more and is enthusiastically looking forward to pursuing a Ph.D. degree. Ahmed can be reached through LinkedIn (linkedin.com/in/AhmedFGad), Facebook (fb.com/AhmedFGadd), and via e-mail (ahmed.f.gad@gmail.com or ahmed.fawzy@ci.menofia.edu.eg).

About the Technical Reviewer

 **Yurii Sekretar** is a professional software developer with over 14 years of experience. He is currently shaping the future of smart intercom at ButterflyMX. There, he leads the mobile team and is working on a mobile app used by people to enter their homes. He also freelances for several companies and clients, which allows him to study how people interact with different platforms, and it helps him keep up to date with the fast-growing technology world. He currently lives in California with his family.

Acknowledgments

Thanks to Allah who planted the idea of starting this book in my mind. Writing a book is not easy but Allah helps me meet the challenges. I always remember Ayah 15 from Surah Al-Naml in the Nobel Quran: "And We had certainly given to David and Solomon knowledge, and they said, 'Praise [is due] to Allah, who has favored us over many of His believing servants.'"

Thanks to the Apress team, including Celestin John, Divya Modi, Welmoed Spahr, and Rita Fernando. This is my second time working with them and I happy to repeat the experience.

I am grateful to the Kivy community, led by Mathieu Virbel and Gabriel Pettier, who did a good job building such a wonderful open source framework. It makes it easier to build cross-platform apps in Python. I implore companies to fund this awesome project.

Thanks to Matthew Mayo, a researcher and editor at KDnuggets, for sharing my tutorials with readers all over the world. Thanks to Austin Kodra, head of the Fritz community, for his trust and for making me part of *Heartbeat* and sharing my writings.

Thanks to my colleague, Mohamed Hamed, a teaching assistant and a senior Android app developer, for giving me Android Studio packaged with the necessary tools to start app development.

I am grateful to Dr. Mahmoud Albawaneh, Executive Director of Institutional Research and Analytics at California State University, Long Beach, for helping me get out of my comfort zone and for always helping me.

I am delighted to say thanks to my friend Fatima Ezzahra Jarmouni, a Moroccan data scientist, for the continuous encouragement and for never letting me walk alone.

I am very thankful to Monica Abdullah Yunis, a kind Palestine-American Muslim mother, for teaching me perseverance through fasting in order to cleanse my soul and mind just for Allah. This is a cure for depression. She is always a wise supporter of me and available to help me.

I can never forget my childhood friends, Abdelmaksoud Abou Koura, Ibrahim Osama, Islam Al Hag Aly, and Mohamed Samy. You and your families played a critical role in building my character and advising me in the transitional moments of my life.

ACKNOWLEDGMENTS

Thanks to my social media followers for being a part of my journey.

Finally, thanks to my family, especially my mother, for being with me always. I ask forgiveness to my father since his death in 2001. I try to do my best to prove that I deserve every breath I take. It is my way of thanking everyone who helped me and proving that they were not mistaken for doing so.

Introduction

Python is a popular programming language that grabs the attention of its users because it's so simple and powerful. A complex task is accomplished in Python using a few lines of code. Python has a number of libraries, such as NumPy, that make life easier. Python is an essential requirement for data scientists.

Once you build a Python desktop app, it's a good idea to think about distributing it to different platforms. The good news is that there is a Python framework called Kivy (`https://github.com//kivy/kivy`) for building cross-platform apps with natural user interfaces. Using the same code unchanged, you can produce apps for the Windows, Linux, Mac, iOS, and Android platforms. This book focuses on building Android apps and exploiting the Android features using Kivy. Multiple user interface elements (widgets) are introduced throughout the book.

The framework is very easy to learn once you understand the basic app structure. Because Kivy is simple, the challenge is not learning the Kivy features but using them effectively to build rich apps. This book provides a good recipe for building a number of apps that effectively use Kivy features. Let's look at an overview of the book chapters.

Chapter 1 introduces Kivy and prepares its development environment in a Linux platform for building Kivy desktop apps. Throughout the book, a desktop Kivy app is created first to check whether everything is working as expected. After that, this chapter produces Android apps. Buildozer and Python-4-Android are the two projects used to produce the Android app. This is by building an Android Studio project based on the Kivy project. The structure of the Android Studio project maps between the Kivy app and the Android app. Once the Android Studio project is created successfully, the APK file of the app can be installed on Android devices. Moreover, the app can be published at Google Play, similar to regular Android apps.

Chapter 2 introduces the KV language for placing the widget tree of the Kivy app in a different file than the Python file. This makes debugging the app much easier.

Chapter 3 discusses the Camera widget, which enables you to access the Android camera very easily. Kivy canvases are discussed for drawing on the Kivy widgets and applying transformations. The captured images are uploaded to a Flask HTTP server to display them in a Web browser.

Chapter 4 uses the Screen widget for separating the widgets across multiple screens, where each screen has a task to do. Navigation from one screen to another is discussed.

Chapters 5 and 6 build a multi-level game from scratch in which a player has a mission of collecting a number of uniformly distributed coins on the screen. Monsters and fires kill the player if a collision occurs.

Chapter 7 is a complete guide for understanding how to edit the Android Studio project produced from the Kivy app within Android Studio. After the project is exported using Buildozer, it is imported in Android Studio as a regular Android app. By doing this, you can add whatever functionality you want to the Android app. If the functionality is not available within the Kivy app, you can add it within Android Studio. This chapter discusses how to access the Kivy widgets within Android Studio to handle their actions. The loading screen of the Kivy app is edited. Moreover, OpenCV is imported into the project and an image is loaded and processed once a Kivy button is pressed.

CHAPTER 1

Preparing Kivy for Android Application Development

This chapter introduces the Kivy framework for building cross-platform applications in Python. It starts by showing you how to prepare the Kivy development environment on a desktop computer. We will create a simple Kivy desktop application. We will also install Buildozer and use it to build an Android Studio project from the desktop app. An APK file is created, which can be installed on an Android device or deployed on Google Play after being signed.

What Is Kivy?

Kivy is a cross-platform Python framework for building applications with a natural user interface (NUI). Being cross-platform, the Kivy code will run unchanged on Windows, Linux, Mac, Android, and IOS devices. The Kivy interface is *natural*, which means it's easy for any user to interact with the interface naturally. Users do not have to spend hours learning how to use the interface. Interaction can be via a cursor with non-touch screens or using multi-touch features.

Kivy is based on a number of low-level libraries that create the user interfaces. In this book, the Simple DirectMedia Layer (SDL) will be used. The SDK is a cross-platform library for low-level access to graphics hardware via OpenGL. Other libraries are also available, such as PyGame. The developer will not interact directly with these libraries because Kivy is abstract. It decouples the developer from unnecessary details and provides a simple interface to access the features of the devices. As an example, in Java,

© Ahmed Fawzy Mohamed Gad 2019
A. F. M. Gad, *Building Android Apps in Python Using Kivy with Android Studio*,
https://doi.org/10.1007/978-1-4842-5031-0_1

1

developers have to write a lot of code in order to access the Android camera and capture an image in Android Studio. In Kivy, many of these details are hidden. In a very easy way, with just a few lines of code, users can capture an image , upload and play an audio file, handle touch events, and more.

Kivy is also modular. This means the Kivy project can be organized into a number of independent modules. Modules can be used in more than one project. In non-modular libraries, there is no separation of the functions being created and thus we have to repeat them for each feature in each new application.

Kivy Installation

In order to build a Kivy mobile application, it is good practice to build a desktop application first. This way, we at least know that things are running well before we move to the mobile app. This is because debugging desktop applications is easier. As a result, this chapter starts by preparing the Kivy development environment on a desktop.

We can install Kivy using the *pip installer,* which fetches the library either from the Python Package Index (PyPI) to install the latest stable version or specifies the link of the GitHub project to install its latest development version. We need to make sure pip is properly installed. If you intend to use Python 2, then type `pip` in the terminal. Use `pip3` for Python 3. Because Python 2 is pre-installed in Ubuntu, it is expected that the `pip` command will succeed. If not, then use the following command:

```
ahmed-gad@ubuntu:-$sudo apt install python-pip
```

If you intend to use Python 3 and the `pip3` terminal command fails, you have to install it using the following command:

```
ahmed-gad@ubuntu:-$sudo apt install python3-pip
```

After installing pip, we can use it to install Kivy. Before installing it, we need to make sure that Cython is installed. Use either `pip` or `pip3` to install Cython if it's not previously installed. Here is the command to install Cython for Python 2:

```
ahmed-gad@ubuntu:-$pip install cython
```

After that, we can start Kivy installation using the following command. Note that it is also installed inside Python 2.

```
ahmed-gad@ubuntu:-$pip install kivy
```

At the current time, we can build Kivy desktop applications before building the mobile applications. Let's start by building a hello world application. The application shows a window that displays the "Hello World" text.

Building a Simple Desktop Kivy Application

The first step in building a Kivy application is creating a new class that extends the kivy.app.App class. According to Listing 1-1, the class is called TestApp. The second step is to override the function called build inside the kivy.app.App class. This function places the widgets (GUI elements) that appear on the screen when running the application. In this example, a label is created from the kivy.uix.label module using the Label class with its text property assigned to the "Hello World" text. Note that the kivy.uix module holds all Kivy widgets that appear on the GUI. Finally, the custom class is instantiated in order to run the application. The complete code for building such an application is shown in Listing 1-1.

Listing 1-1. Build Your First Kivy Application

```
import kivy.app
import kivy.uix.label

class TestApp(kivy.app.App):

    def build(self):
        return kivy.uix.label.Label(text="Hello World")

app = TestApp()
app.run()
```

If the application is saved into a file named test.py, we can run it from the terminal using the following command:

```
ahmed-gad@ubuntu:-$python test.py
```

Figure 1-1 shows the window that appears after running the application. It is just a simple window with text in the center. Note that the word Test is displayed in the title bar of the window. Note that the custom class name is TestApp, which consists of two words, Test and App. When the class has the word App, Kivy automatically sets the application title to the text before it, which is Test in this example.

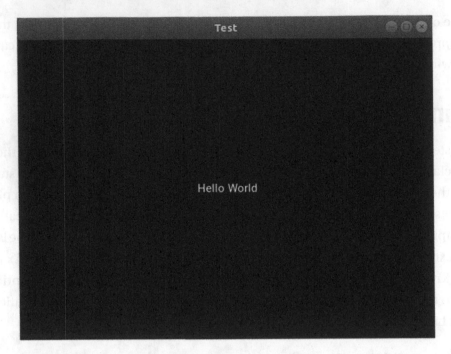

Figure 1-1. *The window of the application in Listing 1-1*

The application title can be changed using the title argument of the `TestApp` class constructor. Listing 1-2 makes two changes to Listing 1-1. The first change is setting the title of the application to `Hello`. The second change is that we can build an application without the `build()` function. By deleting this function, the class will be empty. For such purposes, the `pass` keyword is added.

Listing 1-2. Changing the Application Title

```
import kivy.app

class TestApp(kivy.app.App):
    pass

app = TestApp(title="Hello")
app.run()
```

We currently use this `build()` function to hold the application widgets, so by deleting it, the application window will be empty, as shown in Figure 1-2. In the following chapters, the KV language will be discussed for adding widgets to the application.

Figure 1-2. The window that appears after running the code in Listing 1-2

Installing Buildozer and Creating the buildozer.init File

After creating such a simple desktop application and making sure everything works as expected, we can start building the mobile application. For that, we use a project called *Buildozer,* which is responsible for automating the process of packaging the required tools that make the application run on the mobile end. As stated previously, we can use pip directly to install the latest stable version or we can specify the GitHub link to install the latest development version. We can use pip for its installation according to the following command:

```
ahmed-gad@ubuntu:-$pip install buildozer
```

In order to build the Android application, the Python file must be named `main.py` and be located at the root of the project. Buildozer uses this file as the main activity in the final Java project. After creating this file, we need to specify some properties about the project. These properties are very important to build the Android application. Such properties are added to the file called `buildozer.spec`. You do not need to start this file from scratch, as you can create it automatically using the following command. Just make sure to execute it in the same path where `main.py` is located (i.e., execute it at the project root).

```
ahmed-gad@ubuntu:-$buildozer init
```

Assuming that the folder holding the project files is named NewApp, after issuing this command, the project directory tree will be as follows:

- NewApp
  - bin
  - .buildozer
  - main.py
  - buildozer.spec

The .buildozer folder contains the Android project in addition to its requirements. After we successfully build the project, the bin folder holds the generated APK file.

An abstract version of the buildozer.spec file is given in Listing 1-3. Let's discuss the existing fields, one by one. The title defines the application title, which appears to the user after installing the application. The package.name and package.domain fields are important, as they define the ID of the application when published at Google Play. The source.dir property indicates the location of the Python source file main.py. If it is located in the same folder as the buildozer.spec file, then it's just set to a dot (.).

Listing 1-3. Fields Inside the buildozer.spec File

```
[app]

# (str) Title of your application
title = FirstApp

# (str) Package name
package.name = kivyandroid

# (str) Package domain (needed for android/ios packaging)
package.domain = com.gad

# (str) Source code where the main.py live
source.dir = .

# (list) Source files to include (let empty to include all the files)
source.include_exts = py,png,jpg,kv,atlas,wav

# (list) Source files to exclude (let empty to not exclude anything)
source.exclude_exts = gif
```

```
# (list) List of inclusions using pattern matching
#source.include_patterns = assets/*,images/*.png

# (list) List of exclusions using pattern matching
#source.exclude_patterns = license,images/*/*.jpg

# (list) List of directory to exclude (let empty to not exclude anything)
source.exclude_dirs = bin

# (str) Application versioning (method 1)
version = 0.1

# (list) Application requirements
# comma separated e.g. requirements = sqlite3,kivy
requirements = kivy, numpy

# (str) Custom source folders for requirements
# Sets custom source for any requirements with recipes
requirements.source.numpy = /home/ahmedgad/numpy

# (str) Presplash of the application
presplash.filename = %(source.dir)s/presplash.png

# (str) Icon of the application
icon.filename = %(source.dir)s/logo.png

# (str) Supported orientation (one of landscape, portrait or all)
orientation = landscape

#
# OSX Specific
#

# change the major version of python used by the app
osx.python_version = 3

# Kivy version to use
osx.kivy_version = 1.10.1

#
# Android specific
#
```

```
# (bool) Indicate if the application should be fullscreen or not
fullscreen = 1

# (list) Permissions
android.permissions = INTERNET, CAMERA

# (int) Android API to use
android.api = 26

# (int) Minimum API required
android.minapi = 19

# (int) Android SDK version to use
#android.sdk = 27

# (str) Android NDK version to use
#android.ndk = 18b

# (str) Android NDK directory (if empty, it will be automatically
downloaded.)
android.ndk_path = /home/ahmedgad/.buildozer/android/platform/android-ndk-
r18b

# (str) Android SDK directory (if empty, it will be automatically
downloaded.)
android.sdk_path = /home/ahmedgad/.buildozer/android/platform/android-sdk-
linux
```

Assume that we used some resources in the project that must be packaged within the Android application. This can be done in different ways. Despite being simple, it may waste hours of debugging if it's not done successfully.

One simple way is to specify the extension of such resources inside the source. include_exts property. For example, if all files with the .png, .jpg, and .wav extensions are to be packaged within the application, then the property will be as given here. If that field is empty, then all files within the root directory will be included.

```
source.include_exts = png, jpg, wav
```

In contrast to the `source.include_exts` property, there is a property named `source.exclude_exts` that defines the extensions to be excluded from being packaged. If it's empty, then no files are excluded.

There are also `source.include_patterns` and `source.exclude_patterns` that create patterns to be included and excluded, respectively. Note that they are commented using #.

Similar to `source.exclude_exts`, there is a property called `source.exclude_dirs` that defines the directories to be excluded. For example, the `bin` folder exists but we are not interested in including it. This reduces the size of the APK file.

The `version` property defines the Android app version. When you're uploading a new version of the app to Google Play, this property must be changed to a higher number than the one you used previously.

Inside the `requirements` property, you can declare the required libraries imported within the Python code. For example, if you imported NumPy, then NumPy must be an item in this property. Each requirement will be downloaded the first time it is used.

If you download a requirement and want to use it in the application rather than download a new one, you must define the directory of this requirement within the `requirements.source.<requirement-name>` property after replacing the `<requirement-name>` with the name of the requirement. For example, use `requirements.source.numpy` to define the path for NumPy.

The `icon.filename` property defines the image used as an icon for the application. It can be a PNG file. While the application is loading, there is an image displayed for the user defined by the `presplash.filename` property . The Kivy logo is used as the default image.

The `orientation` property defines the supported orientations of the application. It can be set to `landscape` or `portrait` for using one orientation. To set it according to the device's orientation, set it to `all`.

The `osx.python_version` and `osx.kivy_version` properties define the versions of Python and Kivy, respectively, that are being used.

If the application will run in full-screen mode, then set the `fullscreen` property to 1. This hides the notification bar while the application is running.

The `android.permissions` property sets the required permissions for the application to run. For example, if you access the camera inside the application, then the CAMERA permissions must be declared within that property.

Recently, Google prevented users from uploading an application that targets fewer than 26 APIs. Thus, in order to publish an app to Google Play, the app must target at least 26 APIs. The `android.api` and `android.minapi` fields define the target and the minimum API versions to be used. It is important not to set the `android.api` to a value less than 26.

The `android.sdk` and `android.ndk` fields set the versions of the SDK and NDK used to build the application. If such versions are not available, they are downloaded. You can also download these requirements and specify their path inside the `android.ndk_path` and `android.sdk_path` properties.

There are more fields inside the file that may help you. You can learn more about them by scrolling down inside the `buildozer.spec` file. Their jobs can be deduced from their names. Note that you are not asked to use all the fields in the file. Just use the ones that you need.

Buildozer Templates

Note that using Buildozer for building Android applications is similar to a bridge between Python and Android (i.e., Java). The developer creates a Python project and Buildozer converts it into an Android Java project according to the specifications defined in the `buildozer.spec` file. For such purposes, Buildozer has templates that are filled according to the values used in these properties. Assuming that the package name specified inside the `buildozer.spec` file is `kivyandroid`, you can find the templates inside the directory shown here, given that the project root directory is named `NewApp`.

`NewApp/.buildozer/android/platform/build/dists/kivyandroid/templates`

The template for a file called `x.y` is `x.tmpl.y`. For example, the template for the `AndroidManifest.xml` file is called `AndroidManifest.tmpl.xml`. The template for the `build.gradle` file is called `build.tmpl.gradle`.

build.gradle

The part of the `build.tmpl.gradle` file that is responsible for specifying the target and minimum API is shown in Listing 1-4. The `minSdkVersion` field holds the minimum API level supported by the application. The `targetSdkVersion` field holds the target API level. For `minSdkVersion`, if the variable value `{{args.min_sdk_version}}` is replaced by a static value such as 19, then whatever the value specified in the android.minapi property inside the buildozer.spec file, the minimum API will be 19. This also works for targetSdkVersion.

Listing 1-4. Specifying the Target and Minimum APIs in Gradle

```
android {
    ...
    defaultConfig {
        minSdkVersion {{ args.min_sdk_version }}
        targetSdkVersion {{ android_api }}
        ...
    }
    ...
```

AndroidManifest.xml

Inside the AndroidManifest.templ.xml file, the part responsible for declaring the application permissions defined in the android.permissions property of the buildozer.spec file is given in Listing 1-5. The first line allows writing to external storage. Because such permission is absolute and does not depend on the permissions defined in the buildozer.spec file, that means such permissions exists even if you did not specify any.

Listing 1-5. Declaring the Application Permission Inside Android Manifest

```
    ...
<uses-permission android:name="android.permission.WRITE_EXTERNAL_STORAGE"
/>
{ % for perm in args.permissions %}
{ % if '.' in perm %}
<uses-permission android:name = "{{ perm }}" / >
{ % else %}
<uses-permission android:name = "android.permission.{{ perm }}" />
{ % endif %}
{ % endfor %}
    ...
```

For the custom permissions defined, there is a `for` loop that iterates through each value in the `android.permissions` property. For each value, a `<uses-permission>` element is created for it.

strings.xml

Another template is named `strings.tmpl.xml` and it is responsible for producing the `strings.xml` file that defines the string resources for the application. Listing 1-6 shows the contents of this template. The first string is named `app_name` and it defines the name of the application. The name is retrieved by replacing `{{args.name}}` with the value of the `title` property inside the `buildozer.spec` file.

Listing 1-6. Specifying the String Resources Inside the strings.xml File

```xml
<?xml version="1.0" encoding="utf-8"?>
<resources>
    <string name="app_name">{{ args.name }}</string>
    <string name="private_version">{{ private_version }}</string>
    <string name="presplash_color">{{ args.presplash_color }}</string>
    <string name="urlScheme">{{ url_scheme }}</string>
</resources>
```

After preparing the required files based on such templates, Buildozer builds the Android project. The main activity of the project is `PythonActivity.java` and it is available in the following path. Later in the book, the Android project will be used in Android Studio to write some Java functionalities.

`/NewApp/.buildozer/android/platform/build/dists/kivyandroid/build/src/main/java/org/kivy/android/`

Before building the application, there are some tools that must be available, such as Android SDK and NDK. A simple way to install tools required for building the Android application is using a terminal command like the command shown here. It downloads the SDK and NDK according to the versions defined in the `android.sdk` and `android.ndk` properties in `buildozer.spec`.

`ahmed-gad@ubuntu:-$buildozer android debug`

Buildozer prepares all requirements to guarantee building the application successfully. For example, it packages the required Python libraries into the APK and downloads them if they are not already downloaded. It also fetches other tools, such as SDK, NDK, and Python-4-Android (P4A).

If your Internet connection is fast, this command will make your life easier. But this way is very time consuming for low speed Internet connections and thus not reliable. If the Internet connection is lost, there is no chance to continue the download. Downloading the required Python libraries is not challenging, as their size is not large compared to the SDK and NDK. Based on my experience, I lost a lot of time trying to download such requirements multiple times.

A better solution is to minimize the amount of data being downloaded using Buildozer. This is done by preparing most of the requirements manually (offline) and then linking them to Buildozer. You do this by using more reliable software that supports pausing and resuming the download process. Preparing the requirements offline is interesting due to the ability to select an exact version of whatever you need in the application. An interesting reason for downloading the requirements offline is when we are interested in building more than one project sharing some requirements. The same requirements will be downloaded each time a project is created. Alternatively, we can download the requirements just once and link them to Buildozer. So, we do not need to download the same file for each individual project.

Preparing Android SDK and NDK

Starting with the Android SDK for Linux, it can be download from this page (`http://dl.google.com/android/android-sdk_r{{rev}}-linux.tgz`) after specifying the exact revision number. This URL uses SDK revision 24 (`http://dl.google.com/android/android-sdk_r24-linux.tgz`). It will be downloaded as a compressed file, as the extension reflects.

Because the downloaded SDK comes without some components, including SDK Tools, SDK Platform, SDK Platform Tools, and SDK Build Tools, we need to download them too. The SDK Platform is used to target a specific Android platform. The SDK Platform Tools are used to support the features of the target Android platform. The SDK Build Tools are used for building the application and creating the APK. The SDK Tools provide the tools for development and debugging. For example, after we build the APK file for the target platform, these tools are used to run and debug it.

The SDK Tools can be downloaded from this page (`https://dl-ssl.google.com/android/repository/tools_r{{rev}}-linux.zip`) after specifying the revision number. For example, SDK Tools revision 22.6.2 can be downloaded from this page (`https://dl-ssl.google.com/android/repository/tools_r22.6.2-linux.zip`). The SDK Platform can be downloaded from this page (`https://dl.google.com/android/repository/android-{{platform}}_r{{rev}}.zip`), after specifying the target platform and revision numbers. For example, this is the URL for downloading the SDK Platform 19 revision 4 (`https://dl.google.com/android/repository/android-19_r04.zip`). The SDK Platform tools can be downloaded from this URL (`https://dl.google.com/android/repository/platform-tools_r{{rev}}-linux.zip`), after specifying the revision number. For example, this is the URL to download SDK Platform Tools revision 19 (`https://dl.google.com/android/repository/platform-tools_r19.0.1-linux.zip`).

The SDK Build Tools can be downloaded from this URL (`https://dl.google.com/android/repository/build-tools_r{{rev}}-linux.zip`), after specifying the revision number. This is the URL for SDK Build Tools revision 19.1 (`https://dl.google.com/android/repository/build-tools_r19.1-linux.zip`).

Similar to downloading the SDK, the NDK can be downloaded from this URL (`http://dl.google.com/android/ndk/android-ndk-r{{rev}}c-linux-x86_64.tar.bz2`). This URL corresponds to NDK revision 9 (`http://dl.google.com/android/ndk/android-ndk-r9c-linux-x86_64.tar.bz2`).

The P4A project can be also downloaded automatically and installed using Buildozer. But if we are interested, for example, in using the development version of the project, we cannot depend on Buildozer and have to clone it from GitHub instead (`https://github.com/kivy/python-for-android`).

After successfully downloading the requirements, such as the SDK and its tools, the NDK, and the P4A, there is another step, and that is to link them to Buildozer. Here's how that process works.

Assuming that the installed Buildozer is located in `/home/ahmedgad/.buildozer/`, the compressed Android SDK and NDK will be extracted to the `/home/ahmedgad/.buildozer/android/platform/` path. If the SDK and NDK folders are named `android-sdk-linux` and `android-ndk-r9c`, then the complete path of the SDK is `/home/ahmedgad/.buildozer/android/platform/android-sdk-linux` and the NDK's complete path is `/home/ahmedgad/.buildozer/android/platform/android-ndk-r9c`. Inside the Tools folder of the SDK, the SDK Tools will be extracted. The downloaded SDK Platform will be extracted inside the `platforms` folder, inside the SDK. The SDK Platform Tools will be extracted inside the `platform-tools` folder, inside the SDK. The downloaded SDK Build Tools will be extracted inside the `build-tools` folder, also inside the SDK. The directory tree of the SDK will be as follows, where `...` means there are some files and folders inside the parent:

- `android-sdk-linux`
  - `build-tools`
    - `19.1.0`
      - `...`
  - `platforms`
    - `android-19`
      - `...`
  - `platforms-tools`
    - `...`
  - `tools`
    - `...`
  - `...`

Note that you can manage the SDK using the SDK Manager from the Android tool located in the SDK Tools. It can be accessed using the following command:

```
ahmed-gad@ubuntu:-$. android-sdk-linux/tools/android
```

The Android SDK Manager is shown in Figure 1-3. Using the manager, we can view the already installed tools, available updates, and new tools for download. Make sure the required tools are available.

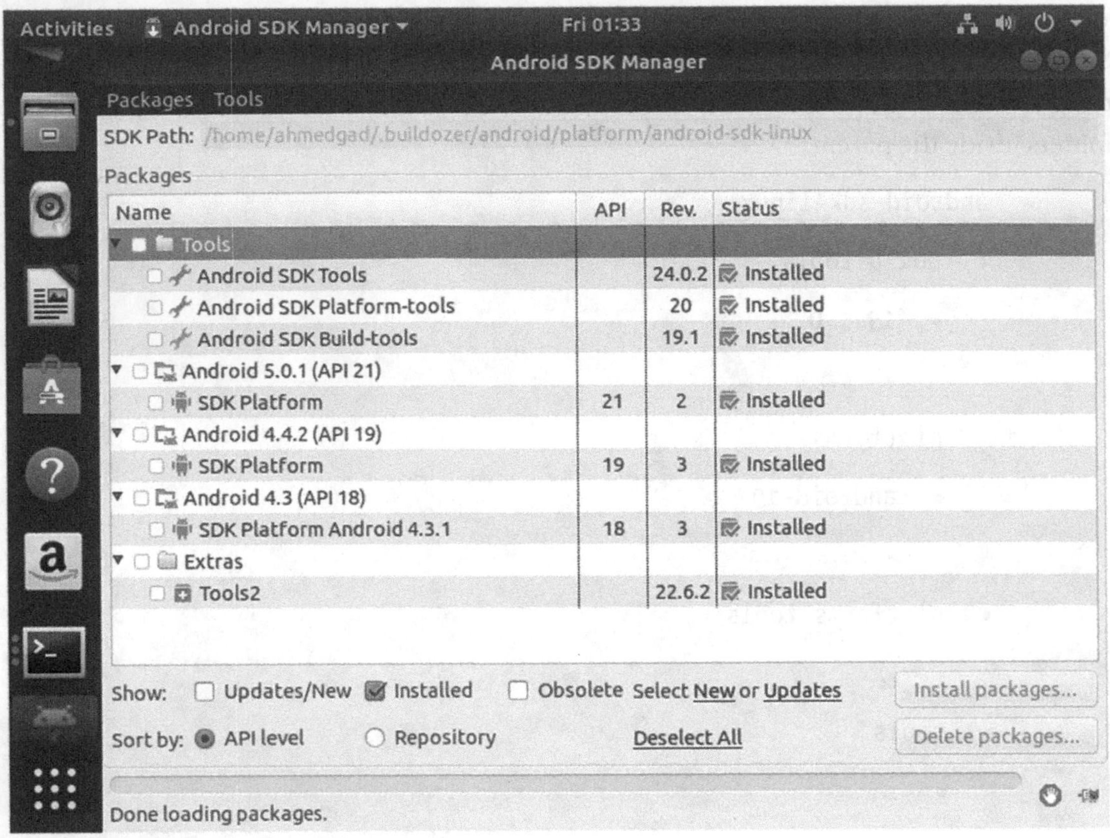

Figure 1-3. *Android SDK Manager*

After that, we need to specify the path of the SDK and NDK inside the `buildozer.` `spec` file of the application using the `android.sdk_path` and `android.ndk_path` fields, respectively. If the path of the cloned P4A project is `/home/ahmedgad/python-for-android`, it will be assigned to the `p4a.source_dir` field inside the `buildozer.spec` file.

Preparing Python Requirements

Let's talk about downloading the required Python libraries by project. As is typical, we do not have to download them offline because Buildozer itself downloads all required Python libraries and links them automatically to the project. It knows which libraries are required based on the `requirements` field of the `buildozer.spec` file. For example, if our project needs NumPy and Kivy, the field will look as follows:

```
requirements = kivy,numpy
```

By just specifying the required libraries in that field, Buildozer will download them for you. For each library, a folder with its name is created inside the /NewApp/. buildozer/android/platform/build/packages path, where NewApp is the root directory of the project. For example, if NumPy is a required library then there will be a folder called numpy in that path. Inside the folder of each library, the library will be downloaded as a compressed file. The idea is to cache the library so that it will be downloaded just one time rather than downloading them each time the project is built. Note that up to this point the library is not installed but just downloaded/cached.

Note that the libraries don't have to exist inside the packages folder. There is another way to tell Buildozer where the required library resides. Assume that a library is downloaded previously outside the packages folder and we just want to tell Buildozer. The path of this library is specified in the buildozer.spec file. The requirements. source.LIB_NAME field accepts the path of such libraries after replacing LIB_NAME with the library name. For example, the path for NumPy is specified in the requirements. source.numpy field.

Assuming that our project needs NumPy 1.15.2, we can download it from this page (https://github.com/numpy/numpy/releases/download/v1.15.2/numpy-1.15.2.zip). If this library is located in /home/ahmedgad/numpy, the required field to be added into the buildozer.spec file to help Buildozer use this library is as follows. This process can be repeated for all types of requirements.

```
requirements.source.numpy = /home/ahmedgad/numpy
```

When Buildozer builds the project, it checks whether the required libraries are cached into the packages folder or its path is explicitly specified using the requirements.source field. If the libraries are found, they will be used. If not, Buildozer has to download them inside the packages folder.

After Buildozer locates all the required libraries, it starts to install them. The libraries are installed to the /NewApp/.buildozer/android/platform/build/build/other_ builds path. Similar to creating a folder to save each downloaded library, there is also a folder inside that path to holds the installed libraries. For example, there will be a folder called numpy to hold the installation of the NumPy library.

Building and Running the Simple Kivy Android Application

After we have installed the required libraries, the `buildozer android debug` command will complete successfully and the APK file will be exported to the project's `bin` directory. We can copy this file to the Android device, install it, and run the application. We can also use this single command to build, deploy, and run the application automatically to the Android device:

```
ahmed-gad@ubuntu:-$buildozer android debug deploy run
```

This requires connecting the Android device to the machine using the USB cable and enabling USB debugging. To enable USB debugging in the device, go to Settings, select the Developer Options item at the end of the list, and toggle the state of the USB Debugging checkbox to on. If the Developer Options item is not visible, go to the About Device item under Settings and tab it seven times. (Note that the steps may change a bit based on the Android OS version.)

We can also view the log of the device by appending `logcat` to this command, as follows:

```
ahmed-gad@ubuntu:-$buildozer android debug deploy run logcat
```

Figure 1-4 shows how the application appears after running it in an Android device.

Figure 1-4. *Running a Kivy application in Android*

Summary

By reaching the end of this chapter, we have successfully created a very simple Android application using the Kivy cross-platform Python library. Remember that the same Python code is used to create both the desktop and mobile applications. We started the application development process by preparing the Kivy development environment on a desktop computer. Then we created a simple application to check whether everything was working as expected. This application simply displays text in a window.

After running the desktop application successfully, we started the mobile (Android) application deployment. It was started by installing Buildozer for producing the Android Studio project for the Kivy application. The Kivy project properties are specified in a file called `buildozer.spec`. These properties include application title, requested permissions, required resources, paths of the Android SDK and NDK, and more. The project includes all resources necessary for Android Studio, such as Gradle, Manifest, Strings, and others. An APK file was created for this project, which can be installed on an Android device or even deployed at Google Play after being signed.

Using KV Language for Separation of Logic and GUI

In the previous chapter, we prepared the Kivy development environment and created a very simple desktop application. Then we installed Buildozer to build mobile applications. As we previously discussed, the tools and requirements for a successful build can be installed automatically using Buildozer or can be downloaded offline and linked. By preparing the Android SDK, NDK, and the required Python libraries, we can successfully build an Android application that uses the same Python code as the desktop application. This is because Kivy is a cross-platform library.

In this chapter, we are going to enrich our application by placing more widgets on the GUI and adding logic that benefits from the data received by such widgets. At the beginning, the GUI widgets will be added using Python code inside the `build()` function of the `kivy.app.App` class. We will create an application with three widgets (button, text input, and label). Once the button is pressed, the data from the text input will be displayed on the text label. As we add more widgets, the Python code will be complex and thus harder to debug. For that reason, we will separate the GUI from the Python logic using the KV language. That way, the Python code will be dedicated just to the logic. Let's get started.

Adding the TextInput Widget to the GUI

In the previous chapter, we created an application with just a single static label widget. In this chapter, we start by adding a new widget called `TextInput`. From its name, it can be easily deduced that this widget allows the users to input text. The `TextInput` widget

© Ahmed Fawzy Mohamed Gad 2019
A. F. M. Gad, *Building Android Apps in Python Using Kivy with Android Studio*,
https://doi.org/10.1007/978-1-4842-5031-0_2

is created by instantiating the `kivy.uix.textinput.TextInput` class. The constructor of this class receives an attribute called `text`, which is default text inside the widget. Listing 2-1 creates a complete application with this widget.

Listing 2-1. Adding a TextInput Widget to the GUI

```
import kivy.app
import kivy.uix.textinput

class TestApp(kivy.app.App):

    def build(self):
        return kivy.uix.textinput.TextInput(text="Hello World")

app = TestApp()
app.run()
```

If you run this Python script, the window shown in Figure 2-1 will be displayed. The entire window is just text input. You can edit the text that's entered.

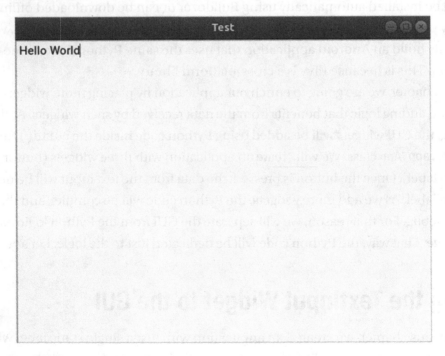

Figure 2-1. *Window of the application created in Listing 2-1*

We can add a button widget to the GUI. By clicking the button, the data from the text input will be received and printed in a `print` statement. There is an issue now. Inside the `build()` function, just a single widget is returned. Now we need to return two widgets (the button and text input). How do we do that? The solution is to add these widgets to a container and then return this container. Containers in Kivy are the layouts.

There are different types of layouts in Kivy, such as box, relative, grid, stack, and more. Each layout has its way of arranging child widgets inside it. For example, the box layout arranges the children vertically or horizontally. The grid layout splits the window into a matrix of rows and columns and then inserts the widgets inside the matrix cells. By default, the window is divided equally across all widgets.

Enriching the GUI Application by Adding More Widgets Inside Python

The code in Listing 2-2 creates an application in which the button and the text input are inserted into a box layout. The GUI of the application is created inside the `build()` function. At first, the button is created as an instance from the `kivy.uix.button.Button` class. It accepts the argument `text`, which accepts the text to be displayed on the button. Both the button and the text input are saved in variables for later use. The button widget is saved into the `my_button` variable, while the text input is saved in the `text_input` variable.

The box layout is created as an instance from the `kivy.uix.boxlayout.BoxLayout` class. It is saved in the `box_layout` variable. To specify whether widgets are added vertically or horizontally inside the layout, the `orientation` argument is specified inside the class constructor. Its default value is `horizontal`, meaning that widgets will be inserted horizontally from left to right. In this example, the orientation is set to vertical, so the widgets will be inserted from top to bottom, where the first element inserted will be placed at the top and the last element inserted will be placed at the bottom of the window. Widgets are inserted into the layout using the `add_widget` function by specifying the name of the widgets to be added. Finally, the layout is returned.

Listing 2-2. Adding More Than One Widget to the Application

```
import kivy.app
import kivy.uix.textinput
import kivy.uix.button
import kivy.uix.boxlayout

class TestApp(kivy.app.App):

    def build(self):
        my_button = kivy.uix.button.Button(text="Click me")
        text_input = kivy.uix.textinput.TextInput(text="Data inside
        TextInput")

        box_layout = kivy.uix.boxlayout.BoxLayout(orientation="vertical")
        box_layout.add_widget(widget=button)
        box_layout.add_widget(widget=textInput)

        return box_layout

app = TestApp()
app.run()
```

After you run the application, the window shown in Figure 2-2 will appear. It is divided vertically into two equal-sized parts and placed according to the order of placement inside the layout.

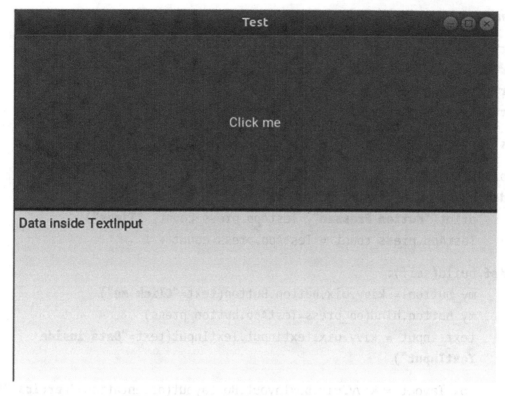

Figure 2-2. *Window of application created in Listing 2-2*

Up to this time, clicking the button causes no action. To handle the button action, the bind() function is used. It accepts an argument reflecting the action. To handle the button press action, use the on_press argument. This argument is assigned to a function that is called when the action is triggered.

Handling a Button Press

The code in Listing 2-3 creates a function named button_press() that is called when the button is pressed. It accepts the widget that fired the action as an argument. This function is attached to the button so that it is executed when the button is pressed. The function prints a message each time the button is pressed showing the number of times the button has been pressed, based on the press_count variable. It is incremented at the end of the function call.

Listing 2-3. Handling a Button Press

```python
import kivy.app
import kivy.uix.textinput
import kivy.uix.button
import kivy.uix.boxlayout

class TestApp(kivy.app.App):

    press_count = 1
    def button_press(self, button_pressed):
        print("Button Pressed", TestApp.press_count, "Times")
        TestApp.press_count = TestApp.press_count + 1

    def build(self):
        my_button = kivy.uix.button.Button(text="Click me")
        my_button.bind(on_press=TestApp.button_press)
        text_input = kivy.uix.textinput.TextInput(text="Data inside
        TextInput")

        box_layout = kivy.uix.boxlayout.BoxLayout(orientation="vertical")
        box_layout.add_widget(widget=my_button)
        box_layout.add_widget(widget=text_input)

        return box_layout

app = TestApp()
app.run()
```

The printed messages after pressing the button four times are shown in Figure 2-3.

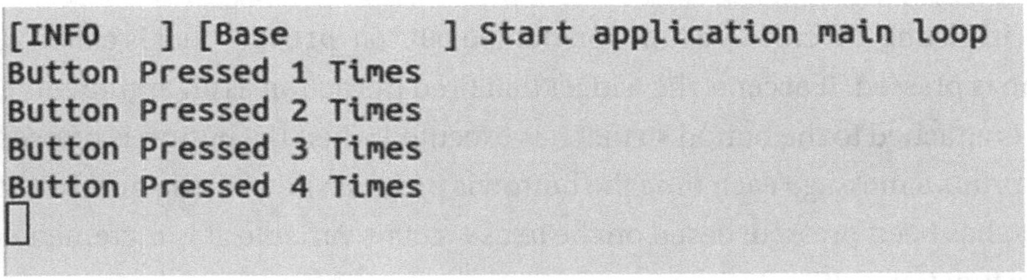

```
[INFO    ] [Base          ] Start application main loop
Button Pressed 1 Times
Button Pressed 2 Times
Button Pressed 3 Times
Button Pressed 4 Times
[]
```

Figure 2-3. *A message is printed each time a button is pressed*

Receiving Data from Text Input

The application could be modified in order to print the text inserted into the TextInput widget. As previously explained, when the button is pressed, its callback function button_press() will be called. Inside this function, the text in the TextInput widget can be returned and printed. In order to be able to access the TextInput widget inside this function, the widget is stored inside the current object referenced by the keyword self. The code of the new application is shown in Listing 2-4.

Listing 2-4. Receiving Text from the TextInput upon a Button Press

```python
import kivy.app
import kivy.uix.textinput
import kivy.uix.button
import kivy.uix.boxlayout

class TestApp(kivy.app.App):

    def button_press(self, button_pressed):
        input_data = self.text_input.text
        print(input_data)

    def build(self):
        my_button = kivy.uix.button.Button(text="Click me")
        my_button.bind(on_press=self.button_press)
        self.text_input = kivy.uix.textinput.TextInput(text="Data inside
        TextInput")

        box_layout = kivy.uix.boxlayout.BoxLayout(orientation="vertical")
        box_layout.add_widget(widget=my_button)
        box_layout.add_widget(widget=self.text_input)

        return box_layout

app = TestApp()
app.run()
```

After the button is pressed, the current text inside the TextInput widget will be fetched using the text property and printed to the terminal.

Displaying Text on a Text Label

At the current state, we have to open the window, press the button, and go to the terminal to see the printed message. If the button is pressed again, we have to go to the terminal to see the output, and so on. We can make the life easier by printing the message on a label widget inside the window itself. As a result, we do not have to open the terminal at all. The changes compared to the last application involve creating a new label widget using the kivy.uix. label.Label class, adding it to the box layout, attaching it to the current object (self) in order to access it within the button_press() function, and changing its text according to the input received from the TextInput widget. The new application is shown in Listing 2-5.

Listing 2-5. Adding a Label Widget to the Application

```python
import kivy.app
import kivy.uix.label
import kivy.uix.textinput
import kivy.uix.button
import kivy.uix.boxlayout

class TestApp(kivy.app.App):

    def button_press(self, button_pressed):
        self.text_label.text = self.text_input.text

    def build(self):
        self.text_input = kivy.uix.textinput.TextInput(text="Data inside
        TextInput")
        my_button = kivy.uix.button.Button(text="Click me")
        my_button.bind(on_press=self.button_press)
        self.text_label = kivy.uix.label.Label(text="Waiting for Button Press")

        box_layout = kivy.uix.boxlayout.BoxLayout(orientation="vertical")
        box_layout.add_widget(widget=self.text_label)
        box_layout.add_widget(widget=my_button)
        box_layout.add_widget(widget=self.text_input)

        return box_layout

app = TestApp()
app.run()
```

The application window is shown in Figure 2-4. When the button is pressed, the text inside the TextInput widget is displayed on the label.

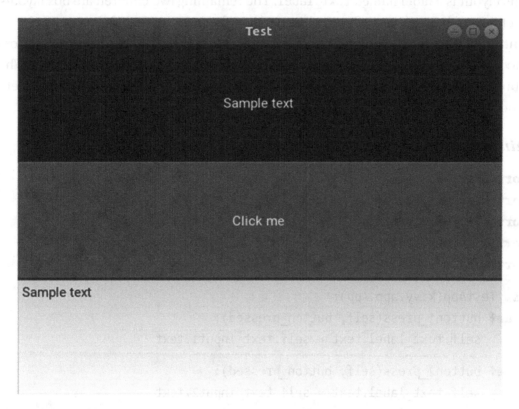

Figure 2-4. *Displaying the text inside the TextInput inside a Label widget after a button press*

Nested Widgets

For the last application, there are just three widgets added as children inside the box layout. Because the orientation of this layout is set to vertical, the window height will be divided equally across the three children, but each child will take the entire width of the window. In other words, each child will take a third of the window height but extend the entire width of the window. It is pretty easy to debug the code in Listing 2-5 because there are just a few widgets. By adding more widgets, the code becomes more complex to debug. We can add more widgets in the next application in Listing 2-6. In the previous application, each widget takes the entire width of the window. In this application, the width of the window is divided across two widgets.

The window will have a root box layout with vertical orientation named box_layout in the application code in Listing 2-6. This layout will have three children. The top child of this layout is a label named text_label. The remaining two children are box layouts (the children themselves are layouts) named box_layout1 and box_layout2. The orientation of each child box layout is horizontal (that is, the children are inserted from left to right). Each child layout will have two child widgets (button and text input). When the button of each child layout is pressed, the text inside the sibling TextInput widget will be displayed on the label.

Listing 2-6. Creating Nested Widgets

```
import kivy.app
import kivy.uix.label
import kivy.uix.textinput
import kivy.uix.button
import kivy.uix.boxlayout

class TestApp(kivy.app.App):
    def button1_press(self, button_pressed):
        self.text_label.text = self.text_input1.text

    def button2_press(self, button_pressed):
        self.text_label.text = self.text_input2.text

    def build(self):
        self.text_label = kivy.uix.label.Label(text="Waiting for Button
        Press")

        self.text_input1 = kivy.uix.textinput.TextInput(text="TextInput 1")
        my_button1 = kivy.uix.button.Button(text="Click me")
        my_button1.bind(on_press=self.button1_press)

        self.text_input2 = kivy.uix.textinput.TextInput(text="TextInput 2")
        my_button2 = kivy.uix.button.Button(text="Click me")
        my_button2.bind(on_press=self.button2_press)
```

```
    box_layout = kivy.uix.boxlayout.BoxLayout(orientation="vertical")

    box_layout1 = kivy.uix.boxlayout.BoxLayout(orientation="horizontal")
    box_layout1.add_widget(widget=self.text_input1)
    box_layout1.add_widget(widget=my_button1)

    box_layout2 = kivy.uix.boxlayout.BoxLayout(orientation="horizontal")
    box_layout2.add_widget(widget=self.text_input2)
    box_layout2.add_widget(widget=my_button2)

    box_layout.add_widget(self.text_label)
    box_layout.add_widget(box_layout1)
    box_layout.add_widget(box_layout2)

    return box_layout
app = TestApp()
app.run()
```

Figure 2-5 shows the window after running the application in Listing 2-6. There is a callback function associated with each button. For example, button1_press() is associated with the first button (my_button1). When a button in a given box layout is pressed, the text from the TextInput widget inside the same box layout is displayed on the label.

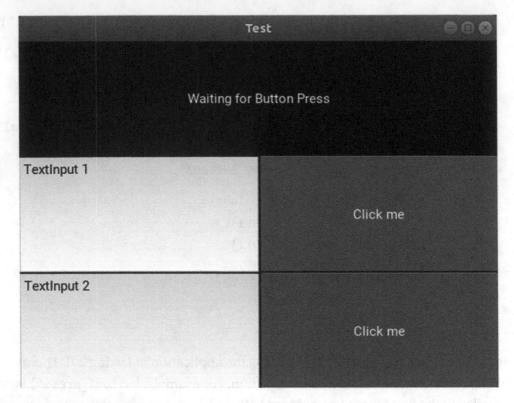

Figure 2-5. *Nested widgets*

After adding more widgets, it is clear that it is difficult to deduce the widget tree of the application. For example, it is not easy to determine the children of a given parent. For that reason, we will use the KV language next, which builds the widget tree of the GUI in a structured way.

Using the KV Language

The KV language (kvlang or Kivy language) creates a widget tree in a readable way that helps us debug the GUI of the application. It uses indentations to mark the children of a given parent. Also it uses indentations to mark the properties of a given widget. Another benefit of using the KV language is the separation of the Python logic from the GUI. The widget tree is created in a file with a `.kv` extension. Thus we can modify the widget tree independently from the Python code. Note that we do not have to import a module into the KV file to use a widget. For example, in order to use a box layout, we just write `BoxLayout`.

Inside the KV file, there is always a widget without any indentation. This is the root widget and it corresponds to the box_layout widget shown in the code in Listing 2-6. The properties and children of this widget are indented four spaces under it. Listing 2-7 shows the contents of the previous application's KV file from Listing 2-6.

Listing 2-7. Using the KV Language to Separate the Python Logic from the GUI

```
BoxLayout:
    orientation: "vertical"
    Label:
        text: "Waiting for Button Press"
        id: text_label
    BoxLayout:
        orientation: "horizontal"
        TextInput:
            text: "TextInput 1"
            id: text_input1
        Button:
            text: "Click me"
            on_press: app.button1_press()
    BoxLayout:
        orientation: "horizontal"
        TextInput:
            text: "TextInput 2"
            id: text_input2
        Button:
            text: "Click me"
            on_press: app.button2_press()
```

The widgets are added to the tree according to the required order to produce the same results given in the previous application. What's worth mentioning is that the fields that need to be referenced in the Python code are given IDs. They are the TextInput and Label widgets. Moreover, the on_press action is attached to the buttons using the on_press property, which is assigned to a function called using the keyword app. This keyword in kvlang refers to the Python file that uses this KV file. As a result, app.button1_press() means call the function named button1_press inside the Python file that is linked to this KV file. The question here is how to link a Python file to a KV file. This is very easy.

33

The class created inside the Python file is named `TestApp`. Kivy extracts the text before the word `App` which is `Test`. After converting the text to lowercase (`Test` becomes `test`), Kivy searches for a KV file named `test.kv` in the same folder of the Python file. If such a file is found, Kivy links it implicitly to the Python file. If it's not found, the application will start, but with a blank window. Note that the `build()` function is deleted. If this function exists in the Python code and Kivy does not locate the KV file, the application will not run.

The Python code, after creating the widget tree in a KV file, is shown in Listing 2-8. Now the Python code is very simple to debug. The application works the same whether the widget tree is created using Python or using the KV language.

One note is how to access a widget created in the KV file inside the Python code. Once the widget is given an ID, you can reference it using the `root.ids` dictionary. The keyword `root` refers to the root box layout widget inside the KV file. By indexing the dictionary by the ID of the required widget, it will get returned and thus we are able to access its properties and override it.

Listing 2-8. Python Code for the Application in Listing 2-6 After Defining the GUI in a KV File

```python
import kivy.app

class TestApp(kivy.app.App):

    def button1_press(self):
        self.root.ids['text_label'].text = self.root.ids['text_input1'].text

    def button2_press(self):
        self.root.ids['text_label'].text = self.root.ids['text_input2'].text

app = TestApp()
app.run()
```

Invoking KV Using load_file( )

Assuming that the KV file is not named `test.kv` but `test1.kv`, Kivy will not be able to implicitly locate the KV file. In that case, we have to explicitly specify the file path inside the `load_file()` function from the `kivy.lang.Builder` class, as shown in Listing 2-9. The result of this function is returned by the `build()` function.

Listing 2-9. Explicitly Specifying the Path of the KV File

```python
import kivy.app
import kivy.lang

class TestApp(kivy.app.App):

    def button1_press(self):
        self.root.ids['text_label'].text = self.root.ids['text_input1'].text

    def button2_press(self):
        self.root.ids['text_label'].text = self.root.ids['text_input2'].text

    def build(self):
        return kivy.lang.Builder.load_file("test1.kv")

app = TestApp()
app.run()
```

Invoking KV Using load_string()

You can also write KV language code inside the Python file using the `load_string()` function inside the `kivy.lang.Builder` class. The code is enclosed between triple quotes, as given in Listing 2-10. Note that this way is not recommended because it does not separate the logic from the visualization.

Listing 2-10. Adding the KV Language Code Within the Python File

```python
import kivy.app
import kivy.lang

class TestApp(kivy.app.App):

    def button1_press(self):
        self.root.ids['text_label'].text = self.root.ids['text_input1'].text

    def button2_press(self):
        self.root.ids['text_label'].text = self.root.ids['text_input2'].text
```

```python
    def build(self):
        return kivy.lang.Builder.load_string(
"""
BoxLayout:
    orientation: "vertical"
    Label:
        text: "Waiting for Button Press"
        id: text_label
    BoxLayout:
        orientation: "horizontal"
        TextInput:
            text: "TextInput 1"
            id: text_input1
        Button:
            text: "Click me"
            on_press: app.button1_press()
    BoxLayout:
        orientation: "horizontal"
        TextInput:
            text: "TextInput 2"
            id: text_input2
        Button:
            text: "Click me"
            on_press: app.button2_press()
""")

app = TestApp()
app.run()
```

Note that this chapter did not create an Android application; the focus was on adding more widgets and structuring the application. Do not worry about building an Android application from what was discussed in this chapter, because it is very simple. Just follow the steps discussed in Chapter 1 for building the APK file using Buildozer after preparing the buildozer.spec file.

Summary

Now that we've reached the end of this chapter, let's quickly recap what was discussed in these first two chapters. In the previous chapter, we prepared the Kivy environment for developing desktop applications. Then we installed Buildozer to develop Android applications. We started off simply, by creating an example in which a single-label widget displayed text. After that, more widgets were added to the application using layouts. The button press action was handled using on_press. Since we nested more widgets inside a widget tree, debugging the application became more difficult. For that reason, the KV language was introduced so we could structure the widget tree and separate the GUI from the Python logic.

In the next chapter, the Camera widget will be introduced so we can access the camera very easily. After making sure the desktop application is created successfully, we will move to building the corresponding Android application and see how accessing the Android Camera widget using Kivy is very intuitive. Because accessing the Android Camera needs permission, the next chapter discusses adding permissions inside the buildozer.spec file. These permissions will be reflected in Google Play for any user to review before installing the application. The next chapter also discusses one of the more important features of Kivy, which is Canvas. Canvas is used for drawing on the widgets and doing transformations.

CHAPTER 3

Sharing Android Camera to an HTTP Server

In the previous two chapters, we prepared the Kivy environment for developing desktop applications. After making sure everything was working as expected, we installed Buildozer to build the Android applications. The first application we created in the book was very simple; a single label widget displayed some text. After that, more widgets were added to the application using layouts. The button press action was handled using on_press. A nested widget tree was created, which made debugging much more complicated. For that reason, the KV language was introduced for structuring the widget tree and separating the GUI from the Python logic.

This chapter discusses accessing and using the Android Camera for capturing and sharing images with an HTTP server. The Camera widget is used to access the Android Camera. After making sure everything is working on the desktop, we use Buildozer to build the Android application. The proper permissions are specified in the buildozer. init file. The Android Camera is, by default, rotated 90 degrees and Kivy canvases are used to handle this issue. Three canvas instances will be discussed—canvas, canvas. before, and canvas.after. In order to limit the effect of a given instruction to just some widgets, the PushMatrix and PopMatrix instructions are discussed.

After the camera is previewed at the proper angle, images will be captured in order to upload them to an HTTP server. The server is created using Flask and runs on a desktop computer. Using the IPv4 address and port number of the server, the requests to the Python library will upload the captured images using the Kivy Android application and an HTTP POST message.

© Ahmed Fawzy Mohamed Gad 2019
A. F. M. Gad, *Building Android Apps in Python Using Kivy with Android Studio*,
https://doi.org/10.1007/978-1-4842-5031-0_3

This chapter ends by creating a live preview of the Android Camera in the web browser of the server. The images are saved into the device memory as a bytes array without being saved in the device storage, in order to save time. Such bytes arrays are then uploaded to the server. The server then interprets these bytes array in order to show the images on an HTML page in the web browser.

Kivy Camera Widget

There are different libraries in Python to access the camera, such as OpenCV and PyGame. Kivy also supports a widget called Camera to access the camera. It is simpler as it does not ask to use a library. The size of the APK file that uses a widget is smaller than the APK in which a library is packaged.

The KV file that has a Camera widget inside the BoxLayout root widget is shown in Listing 3-1. The preferred resolution is specified using the resolution property. Capture images at 1280x720 if possible. Notice the difference between the Camera widget size and resolution. The widget size sets the size of the widget on the application GUI, whereas the resolution defines the pixel density of the captured images.

The play property specifies whether the camera is played after the application starts or not. If it's set to True, the camera will play after the application starts. This simple KV file is the minimum code required to access the camera.

Listing 3-1. Adding the Camera Widget to the Widget Tree

```
BoxLayout:
    Camera:
        resolution: 1280,720
        play: True
```

The Python code associated with this KV file is also very simple and is shown in Listing 3-2. Just create a class that extends the kivy.app.App class and override its build() function. Because this function is empty, the only way to link the Python code to the KV file is to have it named test.kv.

Listing 3-2. Python File Associated with the KV File in Listing 3-1

```python
import kivy.app

class TestApp(kivy.app.App):

    def build(self):
        pass

app = TestApp()
app.run()
```

The application window is shown in Figure 3-1.

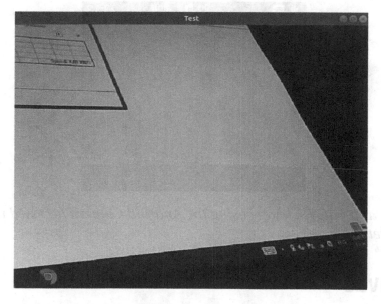

Figure 3-1. *Accessing the camera using the Camera widget*

Accessing Android Camera

At this point, we have created a desktop application that accesses the camera. Let's start building the Android application.

As discussed in Chapter 1, the application's Python file must be named main.py. In order to have permission to access the Android Camera, the android.permissions field must specify such permission as follows:

```
android.permissions=Camera
```

After that, the Android application can be built using Buildozer according to the following command. Remember that this command builds the APK file in debug mode, deploys it to the USB connected Android device, and runs the application after it's installed.

```
ahmed-gad@ubuntu:~$buildozer android debug deploy run
```

Figure 3-2 shows one of the captured images using the Android application. It is rotated counterclockwise 90 degrees. To solve this issue, the widget must be rotated clockwise 90 degrees. Because clockwise rotation uses a negative angle, the rotation by an angle -90 is required. Kivy supports canvases that apply transformations to its widgets.

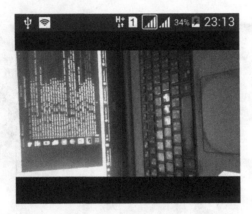

Figure 3-2. *Capturing an image using the Android Camera accessed using the Camera widget*

Kivy Canvas

The area in which we draw is commonly know as the *canvas*. In Kivy, canvases are containers of instructions that define the graphical representation of the widget, not the drawing area. There are three canvas instances in Kivy—canvas, canvas.before, and canvas.after. So, each widget can be assigned these three different instances.

Inside each of the three instances, there are two types of instructions that canvases can do—context and vertex. Vertex instructions draw on the widgets. For example, if a rectangle or a line is to be drawn on a widget, it is a vertex instruction. Context instructions do not draw anything but just change how things appear on the screen. For example, context instructions can transform the widget by changing its rotation, translation, and scale.

Before adding a canvas instruction, a `canvas` instance must be attached to the widget of interest. After that, we can add instructions. For example, the code in Listing 3-3 attaches the `canvas` instance to a `Label` widget and draws a rectangle using the `Rectangle` vertex instruction.

Listing 3-3. Adding Canvas to the Label Widget to Draw a Rectangle

```
BoxLayout:
    Label:
        canvas:
            Rectangle:
                pos: 0,0
                size: 200, 200
```

The rectangle is placed at the pixel (0, 0), which is origin of the Kivy coordinate system corresponding to the bottom-left corner (except for `RelativeLayout` and `ScatterLayout`). The width and height of the rectangle are set to 200 pixels. Thus, the rectangle starts at the bottom-left corner and extends in the horizontal and vertical directions by 200 pixels. Figure 3-3 shows the rectangle.

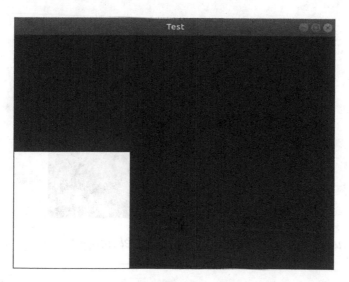

Figure 3-3. *Drawing a rectangle on the Label widget*

We can change the color of the rectangle using the `Color` context instruction. This instruction accepts the RGB color using the `rgb` property, where each channel is given a value between 0 and 1. In Listing 3-4, red is the assigned color.

It is very important to understand that the context instructions are applied to the widgets and vertex instructions below them. If a widget or a vertex instruction is added before the context instruction, the context instruction will not be applied. In this example, if the Color vertex instruction is added after the Rectangle vertex instruction, the rectangle will be colored red.

Listing 3-4. Using the Color Context Instruction to Change the Rectangle Color

```
BoxLayout:
    Label:
        canvas:
            Color:
                rgb: 1, 0, 0
            Rectangle:
                pos: root.pos
                size: 200,200
```

The result, after running the application with the KV file in Listing 3-4, is shown in Figure 3-4. The rectangle is colored according to the Color instruction.

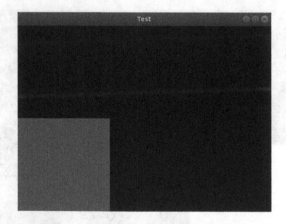

Figure 3-4. *Drawing a red rectangle on the Label widget*

We can repeat the previous instructions for a second label widget according to Listing 3-5. The color applied is green rather than red and the rectangle is positioned at the center of the window.

Listing 3-5. Two Label Widgets Assigned to Two Canvas Instances to Draw Two Rectangles

```
BoxLayout:
    Label:
        canvas:
            Color:
                rgb: 1, 0, 0
            Rectangle:
                pos: root.pos
                size: 200,200
    Label:
        canvas:
            Color:
                rgb: 0, 1, 0
            Rectangle:
                pos: root.width/2-100, root.height/2-100
                size: 200,200
```

The application window in shown in Figure 3-5.

Figure 3-5. *Drawing two rectangles using two Rectangle instructions*

After using the vertex instructions, we can start using the second type of instructions, which is *context*. It is very important to note that the context instructions must be applied before determining where the instructions should be applied. Suppose we rotate a rectangle created using the `Rectangle` vertex instruction. In that case, the context instruction that rotates the rectangle must be added before the `Rectangle` instruction. If the context instruction is added after the `Rectangle` instruction, the rectangle will not be changed. This is because the context instructions are active only before rendering the drawings. The context instructions have no effect after a drawing has been rendered.

The context instruction that rotates the widgets is called `Rotate`. According to Listing 3-6, this context instruction is added before the `Rectangle` vertex instruction to rotate the rectangle. Using the `angle` property of the `Rotate` instruction, the rotation is to -45, which rotates it clockwise. The rotation axis (or axes) can be defined using the `axis` property. The value **0,0,1** means rotation around the Z-axis.

By default, the rotation is relative to the origin of the coordinate system (0, 0). In this example, we are not interested in making the rotation around the (0, 0) point but around the window centerpoint. Using the `origin` property, we can change the rotation origin to the center of the window.

Listing 3-6. Using the Rotation Context Instruction to Rotate the Rectangle

```
BoxLayout:
    Label:
        canvas:
            Color:
                rgb: 1, 0, 0
            Rectangle:
                pos: root.pos
                size: 200,200
    Label:
        canvas:
            Color:
                rgb: 0, 1, 0
            Rotate:
                angle: -45
                axis: 0,0,1
                origin: root.width/2, root.height/2
            Rectangle:
                pos: root.width/2, root.height/2
                size: 200,200
```

Figure 3-6 shows the result after rotating the rectangle.

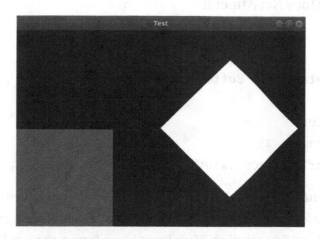

Figure 3-6. *Rotating the rectangle using the Rotation context instruction*

In the previous examples, the context instructions such as `Color` and `Rotate` must be added into the canvas instance before the vertex instructions such as `Rectangle` in order to affect the drawing. The vertex instruction must be written into the KV file on a line before the line in which the target widget is placed. If the widget is located at line 5 for example, then the vertex instruction must be located before line 5, not after it. In the previous examples, we are able to control the placement of the context instructions before the vertex instructions. In some situations, it is not possible to do that.

Let's consider the application shown in Listing 3-7, in which we want to rotate a button.

Listing 3-7. A Button To Be Rotated

```
BoxLayout:
    Button:
        text: "Rotate this Button"
```

If the canvas instance is added inside the `Button` widget according to Listing 3-8, the `Rotate` context instruction inside the canvas instance will be added after, not before, the `Button` widget that we want to rotate. Thus the `Rotate` context instruction will not affect the widget. We need to add the context instruction before, not after, the widget. We will discuss two solutions to this problem.

Listing 3-8. The Rotate Context Instruction Has Been Added After the Button Widget and Thus Does Not Affect It

```
BoxLayout:
    Button:
        text: "Rotate this Button"
        canvas:
            Rotate:
                angle: 45
                origin: root.width/2, root.height/2
```

For a given parent widget, the instructions added to its canvas instance will not only get applied to the parent widget but are also applied to the child widgets. Based on this feature, we can find our first solution. If we want to perform a context instruction on a given widget, we can add that instruction to the canvas instance of its parent. Such instruction will affect both the parent and its child widgets. Listing 3-9 implements this solution. Note that the context instructions not only affect what is drawn using vertex instructions such as Rectangle but also can affect the widgets.

Listing 3-9. Placing the Canvas Instance Inside the Parent Widget in Order to Affect Its Children

```
BoxLayout:
    canvas:
        Rotate:
            angle: 45
            origin: root.width/2, root.height/2
    Button:
        text: "Rotate this Button"
```

The result is shown in Figure 3-7. We solved the problem successfully, but there is an issue.

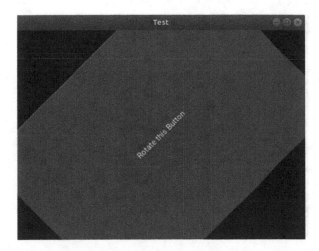

Figure 3-7. *The Button widget is rotated successfully after the canvas is added to its parent*

The previous solution not only rotated the button but also its parent. If there is another child rather than the button, it will be rotated too. The KV file shown in Listing 3-10 has a `Label` widget that should not be rotated. Unfortunately, it will get rotated, as shown in Figure 3-8.

Listing 3-10. Adding the Context Instruction to the Parent Widget Affects All of its Children

```
BoxLayout:
    canvas:
        Rotate:
            angle: 45
            origin: root.width/2, root.height/2
    Label:
        text: "Do not Rotate this Label"
    Button:
        text: "Rotate this Button"
```

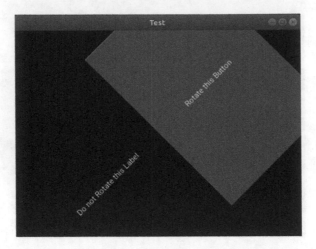

Figure 3-8. *The Rotate context instruction affects the Button and Label widgets*

canvas.before

The previous solution of adding the context instruction to the parent widget affects all child widgets. There is no way to just apply this effect to a specific child. To solve this issue, this solution will use the `canvas.before` instance rather than `canvas` according to the KV file in Listing 3-11. Instructions inside this widget will be executed before the widget is rendered. Thus, if the `Rotate` content instruction is added inside it, the `Button` widget will be rotated successfully.

Listing 3-11. Using canvas.before Rather Than canvas to Rotate the Button Widget

```
BoxLayout:
    Label:
        text: "Do not Rotate this Label"
    Button:
        text: "Rotate this Button"
        canvas.before:
            Rotate:
                angle: 45
                origin: root.width/2, root.height/2
```

The application window is shown in Figure 3-9. Only the `Button` widget is rotated; the `Label` remains unchanged.

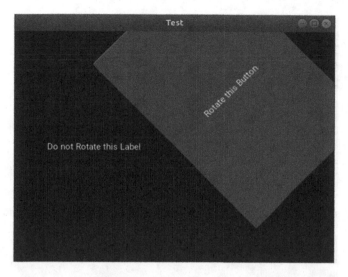

Figure 3-9. *Rotating only one child using the canvas.before instance*

In the previous example, there is a trick. The widget we are looking to rotate is added to the end of the widget tree and this is why the Label is not affected by rotation. If the Label is added after the Button, then the Button and the Label widgets will be rotated. The modified code is shown in Listing 3-12 and the application window is shown in Figure 3-10. Why was the Label widget rotated?

Listing 3-12. Placing the canvas.before Instruction Before the Label Widget

```
BoxLayout:
    Button:
        text: "Rotate this Button"
        canvas.before:
            Rotate:
                angle: 45
                origin: root.width/2, root.height/2
    Label:
        text: "Do not Rotate this Label"
```

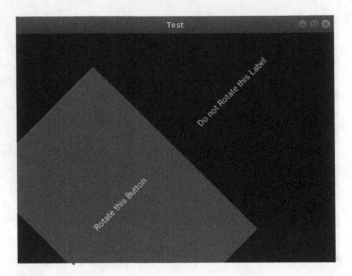

Figure 3-10. *The Button and Label widgets are rotated after canvas.before is added before Label*

The canvas instructions in Kivy are not limited to the widget they are added to. Once an instruction is added in any widget, it will affect other widgets until something cancels the effect of the instruction. For example, if the Button widget is rotated by 45 degrees, then the widgets after it will also be rotated by 45 degrees. If the Label widget comes after the Button and we don't want it to be rotated, we can rotate the Label widget by -45 degrees in order to return it to its original state. The KV file that cancels the Label widget rotation is shown in Listing 3-13. The application window in shown in Figure 3-11. Note that the label is first rotated by 45 degrees then rotated by -45 degrees. If there are more than one widget after the Button, it will be tiresome to rotate them all to return them to their initial states. A better solution is to limit the effect of the Rotate context instruction to just the Button widget.

Listing 3-13. Rotating the Label by -45 Degrees to Cancel the Effect of the Button Rotation

```
BoxLayout:
    Button:
        text: "Rotate this Button"
        canvas.before:
            Rotate:
                angle: 45
                origin: root.width/2, root.height/2
```

```
Label:
    text: "Do not Rotate this Label"
    canvas.before:
        Rotate:
            angle: -45
            origin: root.width/2, root.height/2
```

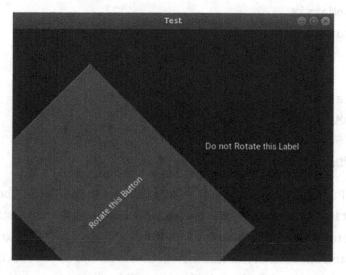

Figure 3-11. *The Left widget is unchanged after rotating it back by -45 degrees*

canvas.after, PushMatrix, and PopMatrix

In order to avoid applying the Rotate instruction to the widgets below the Button widget and limit the effect to the Button widget, Kivy provides the PushMatrix and PopMatrix instructions. The idea is to save the current context states represented by rotation, translation, and scale. After saving the state, we can apply rotation to the Button widget. After the rotated Button widget is rendered successfully, we can restore the saved context state. Thus, only the Button widget will be rotated and all other widgets will retain their context state.

Listing 3-14 shows the KV file that uses PushMatrix and PopMatrix. The resultant window is identical to the one shown in Figure 3-11.

Listing 3-14. Using PushMatrix and PopMatrix to Limit the Effect of the Context Instructions

```
BoxLayout:
    Button:
        text: "Rotate this Button"
        canvas.before:
            PushMatrix:
            Rotate:
                angle: 45
                origin: root.width/2, root.height/2
        canvas.after:
            PopMatrix:
    Label:
        text: "Do not Rotate this Label"
```

Note that the `PushMatrix` instruction is inserted inside the `canavs.before` instance, while the `PopMatrix` instruction is inserted inside the `canvas.after` instance. Adding the `PopMatrix` instruction inside `canvas.after` ensures that it will be executed only after the `Button` is rotated successfully. If this instruction is added to `canvas.before`, then the button will not be rotated. In fact, the button will be rotated according to the `Rotate` instruction and then the context state will be restored before rendering the rotated button. Thus we will not feel by the effect of rotation.

Camera Rotation

After understanding how canvases and their instructions work, we can rotate the `Camera` widget, which is our original target. We can build the application using the KV file shown in Listing 3-15. The file uses the instructions discussed previously (`canvas.before`, `canvas.after`, `PushMatrix`, and `PopMatrix`). It is important to be familiar with them before going further. Note that we are changing the KV file without changing the Python code.

Listing 3-15. The KV File That Rotates the Camera Widget

```
BoxLayout:
    Camera:
        resolution: 1280, 720
        play: True
        canvas.before:
            PushMatrix:
            Rotate:
                angle: -90
                axis: 0,0,1
                origin: root.width/2, root.height/2
        canvas.after:
            PopMatrix:
```

To show the complete idea, the Python code used in Listing 3-2 is repeated in Listing 3-16.

Remember to set the name of the KV file to `test.kv` in order to match the characters before the word App in the class name after converting them to lowercase. Also remember to add CAMERA to the `android.permissions` field of the `buildozer.spec` file to have the permission to use the camera.

Listing 3-16. Python Code Associated with the KV File in Listing 3-15

```
import kivy.app

class TestApp(kivy.app.App):

    def build(self):
        pass

app = TestApp()
app.run()
```

Figure 3-12 shows the application running on an Android device. The widget is placed at the correct angle.

Figure 3-12. *The Camera widget is placed correctly in the Android application*

Before going ahead in this chapter, let's quickly recap what we have discussed so far the book. We prepared the Kivy development environment for building desktop and Android applications. The main tools required were Kivy and Buildozer. We started by creating a simple application in which just a single widget was used. The Kivy BoxLayout container was discussed in order to add more than one Kivy widget to the application window. To allow the user to interact with the application, we discussed how to set and get text widgets such as TextInput and Label. Button press action was handled using on_press.

The Camera widget is also used to access the camera using Kivy. Because the captured images using the Android Camera are by default rotated counterclockwise by 90 degrees, we have to rotate the Camera widget clockwise by -90 degrees. The canvas instance in Kivy allows us to do transformations to the widgets using the context instructions. Moreover, canvases have vertex instructions for drawing shapes on the widgets such as rectangles. Other Canvas instances were discussed, which are canvas. before and canvas.after. In order to limit the effect of the canvas instructions to just a selected widget, the PushMatrix and PopMatrix instructions were discussed.

In the next section of this chapter, we are going to extend the application created previously in order to view the camera, capture an image, and upload it to an HTTP server. We build an application that not just views the camera but also captures images.

An HTTP server is created using the Flask API, which runs on the PC. The server has an open socket based on its IPv4 address and a port number, which waits for requests asking for uploading files. Using the requests Python library, the Kivy application uploads the captured image to the server using an HTTP POST message. Once the server receives an HTTP POST message from the Kivy application, it uploads the file to a selected directory. Let's get started.

Capturing and Saving an Image Using Kivy

We now want to modify the application written in Listings 3-15 and 3-16 to capture and save an image on button press. For this, a Button widget will be added to the end of the window, as shown in Listing 3-17. The question is, how do we capture the camera images? Generally, when a widget calls the export_to_png() function, an image of the widget (i.e., a screenshot) is captured and saved as a PNG file to the specified directory. If the Camera widget called this function, the camera image will be saved as a PNG file.

There are some notes to gain a better understanding of the KV file shown in Listing 3-17. The orientation of the BoxLayout is set to vertical to ensure that widgets are arranged vertically so that the button can be added to the end of the layout. The Camera widget is added before the Button widget so that the button is added to the end of the window.

The Camera widget is given an ID of camera to access it inside the Python code to call the export_to_png() function when the button is pressed. The capture() function inside the Python code will be called when the on_press action is fired.

Listing 3-17. Adding a Button that Captures an Image When Pressed

```
BoxLayout:
    orientation: "vertical"
    Camera:
        id: camera
        size_hint_y: 18
        resolution: (1280, 720)
        play: True
```

```
        canvas.before:
            PushMatrix:
            Rotate:
                angle: -90
                origin: root.width/2, root.height/2
        canvas.after:
            PopMatrix:
    Button:
        text: "Capture"
        size_hint_y: 1
        on_press: app.capture()
```

The last note is that the `size_hint_y` property is added to the `Camera` and the `Button` widgets. The size of the widgets is calculated automatically according to the `BoxLayout`. This property hints to the layout to make the height of a given widget larger, smaller, or equal to the height of another. For example, if the `size_hint_y` is set to 2 for the camera and 1 for the button, then the `Camera` widget height will be double the height of the button. If it is set to 3 for the camera and 1 for the button, then the camera height will be three times the button height. If both are set to the same number, the height of the two widgets will be equal. In this example, we assigned a large value to the `Camera` widget and a small value to the button in order to not hide too much area from the screen.

Similar to the `size_hint_y` property, there is a `size_hint_x` property that controls the width of the widgets.

After preparing the KV file, we need to discuss the Python file shown in Listing 3-18. Note that the class is named `PycamApp`. Thus the KV file should be named `pycam.kv` in order to get it implicitly used. Inside the `capture()` function, the `Camera` widget is fetched into the `camera` variable based on its ID. That variable calls the `export_to_png()` function, which accepts the path in which the captured image is saved. You can change this path to customize the application.

Listing 3-18. Python File Associated with KV File in Listing 3-17 that Captures an Image on Button Press

```
from kivy.app import App

class PycamApp(App):

    def capture(self):
```

```
        camera = self.root.ids["camera"]
        camera.export_to_png("/storage/emulated/0/captured_image_kivy.png")
    def build(self):
        pass

app = PycamApp()
app.run()
```

After building and running the Android application, its window is shown in Figure 3-13. By pressing the button, the camera image will be saved in the specified directory. This image is what will be sent to the server. So, let's start building the server.

Figure 3-13. *Capturing an image on button press*

Building the HTTP Server Using Flask

After saving the captured image, we send it to the HTTP server. The server is created using the Flask API, which runs on a PC. Because Flask is out of scope of this book, we will not discuss it in detail. You can read more about Flask in Chapter 7 of my book, *Practical Computer Vision Applications Using Deep Learning with CNN* (Apress, 2018).

Listing 3-19 lists the Flask code for building the HTTP server. The server is made to listen for file upload requests. First, an application instance is created using the flask. Flask class. The constructor of this class accepts the import_name argument, which is set to the folder name that contains the Flask Python file.

Listing 3-19. Building the HTTP Server Using Flask

```python
import flask
import werkzeug

app = flask.Flask(import_name="FlaskUpload")

@app.route('/', methods = ['POST'])
def upload_file():
    file_to_upload = flask.request.files['media']
    file_to_upload.save(werkzeug.secure_filename(file_to_upload.filename))
    print('File Uploaded Successfully.')
    return 'SUCCESS'

app.run(host="192.168.43.231", port=6666, debug=True)
```

At the end of this code, the application listens for requests using the IPv4 address 192.168.43.231 and port number 6666.

If you do not know your IPv4 address, use the ifconfig terminal command. If the command is not found, install net-tools using this command:

```
ahmed-gad@ubuntu:-$sudo apt install net-tools
```

After that, you can execute the ifconfig command, as shown in Figure 3-14.

```
ahmedgad@ubuntu:~/Desktop/FlaskUpload$ ifconfig
ens33: flags=4163<UP,BROADCAST,RUNNING,MULTICAST>  mtu 1500
        inet 192.168.43.231  netmask 255.255.255.0  broadcast 192.168.43.255
        inet6                           prefixlen 64  scopeid 0x20<link>
        ether 00:0c:29:80:ef:ec  txqueuelen 1000  (Ethernet)
        RX packets 99684  bytes 118440198 (118.4 MB)
        RX errors 0  dropped 0  overruns 0  frame 0
        TX packets 46937  bytes 3326509 (3.3 MB)
        TX errors 0  dropped 0 overruns 0  carrier 0  collisions 0

lo: flags=73<UP,LOOPBACK,RUNNING>  mtu 65536
        inet 127.0.0.1  netmask 255.0.0.0
        inet6 ::1  prefixlen 128  scopeid 0x10<host>
        loop  txqueuelen 1000  (Local Loopback)
        RX packets 97909  bytes 84597021 (84.5 MB)
        RX errors 0  dropped 0  overruns 0  frame 0
        TX packets 97909  bytes 84597021 (84.5 MB)
        TX errors 0  dropped 0 overruns 0  carrier 0  collisions 0
```

Figure 3-14. *Using the ifconfig terminal command to determine the IPv4 address*

The debug argument controls whether to activate debug mode. When the debug mode is on, changes to the server will be applied automatically without needing to restart it.

The route() decorator tells Flask which function to execute when a URL is visited. The URL is added as an argument to the decorator and the function is added below the decorator, which is upload_file() in this example. The URL is set to /, which means the root of the server. Thus, when the user visits http://192.168.43.231/:6666, the route() decorator associated with the root directory of the server will receive this request and execute the upload_file() function.

The route() decorator accepts an argument called methods, which receives the type of HTTP messages that the function will respond to as a list. Because we are just interested in the POST HTTP message, it will be set to ['POST'].

The data to be uploaded is sent as a dictionary of key and value. The key is an ID for the file and the value is the file itself. Inside the upload_file() function, the file to be uploaded is fetched using the flask.request.files dictionary. It receives a key referring to the file to be uploaded. The used key is media. This means when we prepare the Kivy application, the key of the image to get uploaded must be set to media. The file is returned into a variable, which is file_to_upload in our example.

If we are interested in saving the file according to its original name, its name is returned using the filename property. Because some files are named to deceive the server and perform illegal actions, the secure filename is returned using the werkzeug. secure_filename() function. After returning the secure filename, the file is saved using the save() function. When the file is saved successfully, a print message appears on the console and the server responds with the word SUCCESS to the client.

Note that the server accepts any file extension. You can read more about Flask to learn how to upload files with specific extensions.

Using Requests for Uploading Files to the HTTP Server

Before modifying the Android application to upload the file, we can build a client as a desktop application that uses the requests library for uploading a file. It will be easier to debug than the mobile application. The application does not include Kivy code and thus we will interact with the server using terminal commands. The client-side code is shown in Listing 3-20.

Listing 3-20. Using Requests to Upload the Captured Image to the Server

```
import requests

files = {'media': open('/home/ahmedgad/Pictures/test.png', 'rb')}
try:
    requests.post('http://192.168.43.231:6666/', files=files)
except requests.exceptions.ConnectionError:
    print("Connection Error! Make Sure Server is Active.")
```

The dictionary holding the files to be uploaded is prepared. It has a single key-value pair. As we stated earlier, the key will be set to media because the server is waiting for that key. The file is opened using the open() function that receives its path. The rb argument specifies that the file is opened as read only in binary mode.

The requests.post() function receives two arguments. The first is the URL that directs it to the root of the server after specifying its socket details (IPv4 address and port). The second one is the dictionary. A try and catch statement is used to check if there is a problem with connection establishment. This ensures that the application will not crash even if there is an error in establishing the connection.

After building both the server and the client, we can start running the server. There is no GUI for the server and thus its interface is the terminal. Because the server is a regular Python file, we can run it by typing the Python filename after the keyword python or python3 according to the Python version used. If its name is FlaskServer.py, it will be executed using the following command. Remember to use the proper path to locate the Python file.

ahmed-gad@ubuntu:~/Desktop/FlaskUpload$python3 FlaskServer.py

We will see informational messages indicating that the server is running successfully, as shown in Figure 3-15.

```
ahmedgad@ubuntu:~/Desktop/FlaskUpload$ python3 FlaskServer.py
 * Serving Flask app "FlaskUpload" (lazy loading)
 * Environment: production
   WARNING: Do not use the development server in a production environment.
   Use a production WSGI server instead.
 * Debug mode: on
 * Running on http://192.168.43.231:6666/ (Press CTRL+C to quit)
 * Restarting with stat
 * Debugger is active!
 * Debugger PIN: 104-290-957
```

Figure 3-15. *Running the Flask HTTP Server*

After running the client-side application and pressing the button, the server receives an HTTP POST message and uploads the file successfully according to the messages printed on the console in Figure 3-16. The HTTP status code of the response is 200 which means the request is fulfilled successfully.

```
ahmedgad@ubuntu:~/Desktop/FlaskUpload$ python3 FlaskServer.py
 * Serving Flask app "FlaskUpload" (lazy loading)
 * Environment: production
   WARNING: Do not use the development server in a production environment.
   Use a production WSGI server instead.
 * Debug mode: on
 * Running on http://192.168.43.231:6666/ (Press CTRL+C to quit)
 * Restarting with stat
 * Debugger is active!
 * Debugger PIN: 104-290-957
File Uploaded Successfully.
192.168.43.231 - - [14/Nov/2018 07:39:18] "POST / HTTP/1.1" 200 -
```

Figure 3-16. *Captured image by the Kivy application is uploaded successfully to the Flask HTTP Server*

Upload the Camera Captured Image Using the Kivy Android Application

After making sure that the desktop client-side application works well, we can prepare the Android application. Compared to Listing 3-18, the changes will be inside the `capture()` function, as shown in Listing 3-21.

Listing 3-21. Capturing and Uploading Images to the Server Using the Android Application

```python
import kivy.app
import requests

class PycamApp(kivy.app.App):

    def capture(self):
        camera = self.root.ids['camera']
        im_path = '/storage/emulated/0/'
        im_name = 'captured_image_kivy.png'
        camera.export_to_png(im_path+im_name)
        files = {'media': open(im_path+im_name, 'rb')}

        try:
            self.root.ids['capture'].text = "Trying to Establish a
            Connection..."
            requests.post('http://192.168.43.231:6666/', files=files)
            self.root.ids['capture'].text = "Capture Again!"
        except requests.exceptions.ConnectionError:
            self.root.ids['capture'].text = "Connection Error! Make Sure
            Server is Active."

    def build(self):
        pass

app = PycamApp()
app.run()
```

Because the server might be offline, we need to reflect that to user. After clicking the Capture button, a message should be displayed informing the user that the connection is being established. If there is a connection error, a message is also displayed to reflect that.

In order to be able to change the text of the button, we have to access it within the Python file. In order to access a widget, it must have an ID. The previous KV file in Listing 3-17 does not assign an ID to the button. The modified KV file in Listing 3-22 assigns an ID to the Button widget. The assigned ID is capture.

Listing 3-22. Assigning an ID to the Button to Access it Inside the Python File

```
BoxLayout:
    orientation: "vertical"
    id: root_widget
    Camera:
        id: camera
        size_hint_y: 18
        resolution: (1280, 720)
        play: True
        canvas.before:
            PushMatrix:
            Rotate:
                angle: -90
                origin: root.width/2, root.height/2
        canvas.after:
            PopMatrix:
    Button:
        id: capture
        text: "Capture"
        size_hint_y: 1
        on_press: app.capture()
```

Because the widgets arrangement did not change compared to the previous application, the application window will be identical to the previous application.

Dynamic IP Address in Requests

Now, the application depends on a static IPv4 address for the server. If the server uses the dynamic host configuration protocol (DHCP), the IPv4 address of the server might change and thus we have to rebuild the application with the new address. To make the process dynamic, we can use a TextInput widget in which the IPv4 address of the server can be entered. Before posting the dictionary to the server, the text from the widget is fetched in order to build the URL. The modified KV file is shown in Listing 3-23. To access this widget, it is assigned the ID ip_address.

Listing 3-23. Adding a TextInput Widget to Enter the IPv4 Address of the Server

```
BoxLayout:
    orientation: "vertical"
    id: root_widget
    Camera:
        id: camera
        size_hint_y: 18
        resolution: (1280, 720)
        play: True
        canvas.before:
            PushMatrix:
            Rotate:
                angle: -90
                origin: root.width/2, root.height/2
        canvas.after:
            PopMatrix:
    TextInput:
        text: "192.168.43.231"
        id: ip_address
        size_hint_y: 1
    Button:
        id: capture
        text: "Capture"
        size_hint_y: 1
        on_press: app.capture()
```

The Python code after using the TextInput widget is shown in Listing 3-24.

Listing 3-24. Fetching the IPv4 Address from the TextInput Widget

```python
import kivy.app
import requests

class PycamApp(kivy.app.App):

    def capture(self):
        camera = self.root.ids['camera']
        im_path = '/storage/emulated/0/'
        im_name = 'captured_image_kivy.png'
        camera.export_to_png(im_path+im_name)

        ip_addr = self.root.ids['ip_address'].text
        url = 'http://'+ip_addr+':6666/'
        files = {'media': open(im_path+im_name, 'rb')}

        try:
            self.root.ids['capture'].text = "Trying to Establish a
            Connection..."
            requests.post(url, files=files)
            self.root.ids['capture'].text = "Capture Again!"
        except requests.exceptions.ConnectionError:
            self.root.ids['capture'].text = "Connection Error! Make Sure
            Server is Active."

    def build(self):
        pass

app = PycamApp()
app.run()
```

The application window is shown in Figure 3-17. Note that the text of the button is changed to "Capture Again!" which means the file is uploaded successfully according to the Python code. Remember to run the server before pressing the button. Try to use a different IPv4 address and note how the button text changes to reflect that there is a connection error.

Figure 3-17. *The image capture application after using the TextInput widget to enter the IPv4 address of the server*

The Flask server log messages also appear on the terminal, as shown in Figure 3-18. The messages reflect that the server received a message of type POST from a client with the IPv4 address 192.168.43.1.

```
ahmedgad@ubuntu:~/Desktop/FlaskUpload$ python3 FlaskServer.py
 * Serving Flask app "FlaskUpload" (lazy loading)
 * Environment: production
   WARNING: Do not use the development server in a production environment.
   Use a production WSGI server instead.
 * Debug mode: on
 * Running on http://192.168.43.231:6666/ (Press CTRL+C to quit)
 * Restarting with stat
 * Debugger is active!
 * Debugger PIN: 104-290-957
File Uploaded Successfully.
192.168.43.1 - - [14/Nov/2018 08:19:26] "POST / HTTP/1.1" 200 -
```

Figure 3-18. *Log messages from the server after receiving a file*

At this point, we successfully created a Kivy Android application that shares a single image captured using the camera to an HTTP server. This is by saving the image to the device storage and then uploading it.

The remaining part of this chapter extends this application by live previewing the Android Camera in the server's web browser. This is done by capturing the camera image and storing it in the device memory as a bytes array. The image will not be stored in the device storage as a file in order to save time. The bytes array is uploaded to the Flask server using an HTML POST message, as was done previously in this chapter. The server receives each image and displays it on an HTML page using the `<img>` element. We create a live Android Camera preview by continuously capturing images, uploading them to the server, and updating and refreshing the HTML page.

Capturing and Storing a Camera Image to the Memory

Previously in this chapter, the Camera widget images are captured and saved as PNG files using the Kivy `export_to_png()` function. When this function is called by a widget, a screenshot from the widget is taken and saved as a PNG file to the directory specified in the function. The saved PNG file is uploaded to the server using the HTTP POST message.

If we would like to continuously preview the captured images using the Android Camera, we have to capture the Camera widget, save each captured image as a PNG file, and post them to the server. Saving a file each time an image is captured is time consuming. Moreover, it is not required to save the file in the storage. We just need to send the captured image to the server as soon as it is available. There's no need to delay the transmission process. For such reasons, we can use tools that capture an image and save the pixels in memory rather than saving it in a file. We can use the `glReadPixels()` function in OpenGL or return the texture of the widget using the `get_region()` function. Both functions capture and save the image to the device memory rather than saving it as a file. This speeds up the process.

The `glReadPixels()` function is available in the `kivy.graphics.opengl` module. Its signature is as follows:

```
kivy.graphics.opengl.glReadPixels(x, y, width, height, format, type)
```

The function works by capturing an image of a region from the Kivy application window. The region starts from the lower-left corner located using the `x` and `y` arguments. The width and height of the region are specified using the `width` and `height` arguments. Using the first four arguments, the region is successfully specified. The region starts at (`x, y`) and extends horizontally to the left by a value equal to `width` and goes vertically up by a value equal to `height`. Because we are interested in capturing the region at which the `Camera` widget is placed, we can return the coordinates of this widget and assign them to the four arguments.

Before returning the pixels of that region, some other arguments need to be specified to control how the pixels are saved into the memory.

The `format` argument specifies the format of the pixel data. There are different values for it, such as `GL_RED`, `GL_GREEN`, `GL_BLUE`, `GL_RGB`, `GL_RGBA`, and more. These values exist in the `kivy.graphics.opengl` module. We are interested in capturing an RGB image and thus will use the `GL_RGB` values.

The `type` argument specifies the type of the pixel data. It has different values such as `GL_UNSIGNED_BYTE`, `GL_BYTE`, `GL_UNSIGNED_SHORT`, `GL_SHORT`, `GL_UNSIGNED_INT`, and more. These values exist in the `kivy.graphics.opengl` module. We are interested in saving the image as a byte array and thus the `GL_UNSIGNED_BYTE` argument is used.

The function has other two optional arguments called `array` and `outputType` that we are not interested in editing.

Another way to capture the image of a widget is using the `get_region()` method of the `kivy.graphics.texture.Texture` class. For any widget with a texture, we can call this function to return its texture. It has four arguments as shown here. They are identical to the first four arguments in the `glReadPixels()` function.

```
get_region(x, y, width, height)
```

You can use any of these functions to capture images from the Camera widget and save the result in the memory. For get_region(), it works with widgets having texture. Some widgets do not have texture such as TextInput and thus we cannot use get_region() with it. On the other hand, glReadPixels() captures images without caring whether a widget has texture or not.

For making things simpler, we can use get_region(). The complete code used to capture an image using get_region() and save it to the memory is shown in Listing 3-25.

We will start building a desktop application to make debugging easier. At the end of the chapter, we can build the Android application.

The Camera widget is fetched using its ID from the KF file to determine its left-bottom location (camera.x, camera.y). This is in addition to its resolution (camera.resolution) to return the captured images size where camera.resolution[0] is the width and camera.resolution[1] is the height. These four values are assigned to the four arguments in the get_region() method.

The get_region() method returns an instance of the TextureRegion class. In order to return the pixels of the texture, we can use the pixels property. It returns the texture of the widget as pixels in RGBA format as an unsigned bytes array. This array is saved in memory. In this example, the bytes array is saved in the pixels_data variable. The data within this variable will later be posted to the server.

Listing 3-25. Capturing a Camera Image Using get_region()

```
import kivy.app
import PIL.Image

class PycamApp(kivy.app.App):

    def capture(self):
        camera = self.root.ids['camera']
        print(camera.x, camera.y)
```

```python
        pixels_data = camera.texture.get_region(x=camera.x, y=camera.y,
        width=camera.resolution[0], height=camera.resolution[1]).pixels

        image = PIL.Image.frombytes(mode="RGBA",size=(int(camera.
        resolution[0]), int(camera.resolution[1])), data=pixels_data)
        image.save('out.png')

    def build(self):
        pass

app = PycamApp()
app.run()
```

Now, we can debug the application to ensure everything works as expected. This is by saving the captured image using Python Image Library (**PIL**). Because `get_region().pixels` returns a bytes array in RGBA format, we need to construct the image from that array. The `frombytes()` function in PIL supports building an image from a bytes array. The function accepts the mode argument by specifying the image mode, which is the string "RGBA" in this example. We also specify IMAGE size in the `size` argument as a tuple and the raw byte data in the **data** argument.

Note that this function accepts the size as an integer. It is better to convert the width and height of the `Camera` widget to integers. This is because the returned width and height of the `Camera` widget might be `float`. The image returned by the `frombytes()` function is saved using the `save()` function.

The KV file used with the previous Python code to build the desktop application is shown in Listing 3-26. It is the same file used in the last application in the previous example except for removing the `TextInput` widget, as we are not interested in contacting the server at the current time.

Listing 3-26. KV File for the Application in Listing 3-25

```
BoxLayout:
    orientation: "vertical"
    Camera:
        id: camera
        size_hint_y: 18
        resolution: (1280, 720)
        play: True
```

```
Button:
    id: capture
    text: "Capture"
    size_hint_y: 1
    on_press: app.capture()
```

After running the application, the window is as shown in Figure 3-19.

Figure 3-19. *An image captured using get_region() and saved into the device memory*

If you click on the button, the get_region().pixels will capture the area of the Camera widget and save it in memory.

Up to this point, we have successfully captured an image using the get_region() method and saved it to device memory. The next step is to send this image to the Flask server.

Posting the Captured Image to the Flask Server Using a HTTP POST Message

In the previous example, the image is saved into the memory as a bytes array and is ready to be sent to the server. Listing 3-27 shows the Python code that sends the array to the Flask server.

Listing 3-27. Uploading the Bytes Array to the Server

```python
import kivy.app
import requests

class PycamApp(kivy.app.App):

    def capture(self):
        camera = self.root.ids['camera']
        print(camera.x, camera.y)

        pixels_data = camera.texture.get_region(x=camera.x, y=camera.y,
        width=camera.resolution[0], height=camera.resolution[1]).pixels

        ip_addr = self.root.ids['ip_address'].text
        url = 'http://'+ip_addr+':6666/'
        files = {'media': pixels_data}

        try:
            self.root.ids['capture'].text = "Trying to Establish a
            Connection..."
            requests.post(url, files=files)
            self.root.ids['capture'].text = "Capture Again!"
        except requests.exceptions.ConnectionError:
            self.root.ids['capture'].text = "Connection Error! Make Sure
            Server is Active."

    def build(self):
        pass

app = PycamApp()
app.run()
```

The returned array of bytes from the get_region() method is inserted into the dictionary that will be sent to the server. Note that the dictionary is assigned to the files argument of the requests.post() function. This means that the bytes array will be received at the server as a file.

Everything else in the code works as discussed in the previous example. Note that we are not interested in using the PIL anymore on the client-side.

The KV file for the client-side application is shown in Listing 3-28, after adding the TextInput widget.

```
BoxLayout:
    orientation: "vertical"
    Camera:
        id: camera
        size_hint_y: 18
        resolution: (1280, 720)
        play: True
    TextInput:
        text: "192.168.43.231"
        id: ip_address
        size_hint_y: 1
    Button:
        id: capture
        text: "Capture"
        size_hint_y: 1
        on_press: app.capture()
```

Before Sending Images to the Server

At the server-side, we are going to receive the uploaded bytes array as a file. This file will be read in order to convert its content to an image using the PIL.Image.frombytes() function, as seen previously. In order to convert the byte array into an image using this function, it receives the size of the returned image in the size argument. Using different size rather than the correct one may degrade the image quality. Thus, we need to know the image size on the server-side. How do we do that?

Every POST message from the client to the server holds the file to be uploaded. We can also send the image size in that message. Unfortunately, this sends more data in every message as the image size is sent every time an image is uploaded. Because the image size is fixed, we do not have to send it more than once.

A better solution is to send a POST HTTP message to the server before sending any images. This message tells the server about the size of the images it will receive in the next messages. When the server receives the uploaded images in new messages, it can use the previously received image size. For such reasons, a new Button widget is added to the end of the widget tree. When pressed, the size of the Camera widget will be fetched and uploaded to the server in a POST HTTP message.

Listing 3-28 shows the modified KV file of the client-side Kivy application. The new button is assigned the ID cam_size. When this button is pressed, the cam_size() function in the Python code will be executed.

Listing 3-28. KV File for the Client-Side Application

```
BoxLayout:
    orientation: "vertical"
    Camera:
        id: camera
        size_hint_y: 18
        resolution: (1280, 720)
        play: True
    TextInput:
        text: "192.168.43.231"
        id: ip_address
        size_hint_y: 1
    Button:
        id: capture
        text: "Capture"
        size_hint_y: 1
        on_press: app.capture()
    Button:
        id: cam_size
        text: "Configure Server"
        size_hint_y: 1
        on_press: app.cam_size()
```

The Python code of the client-side Kivy application is shown in Listing 3-29 after adding the cam_size() function. A dictionary is created to hold the width and height of the images to be uploaded (i.e., the Camera widget). Such data will be sent to the server in the /camSize directory as parameters and thus the params argument of the requests.

post() function is used. If the message is sent successfully to the server, then the newly added Button widget is useless. Thus it will be deleted from the widget tree using the delete_widget() function.

Listing 3-29. Informing the Server by the Width and Height of the Captured Images

```python
import kivy.app
import requests

class PycamApp(kivy.app.App):
    def cam_size(self):
        camera = self.root.ids['camera']
        cam_width_height = {'width': camera.resolution[0], 'height':
        camera.resolution[1]}

        ip_addr = self.root.ids['ip_address'].text
        url = 'http://'+ip_addr+':6666/camSize'

        try:
            self.root.ids['cam_size'].text = "Trying to Establish a
            Connection..."
            requests.post(url, params=cam_width_height)
            self.root.ids['cam_size'].text = "Done."
            self.root.remove_widget(self.root.ids['cam_size'])
        except requests.exceptions.ConnectionError:
            self.root.ids['cam_size'].text = "Connection Error! Make Sure
            Server is Active."

    def capture(self):
        camera = self.root.ids['camera']
        print(camera.x, camera.y)

        pixels_data = camera.texture.get_region(x=camera.x, y=camera.y,
        width=camera.resolution[0], height=camera.resolution[1]).pixels

        ip_addr = self.root.ids['ip_address'].text
        url = 'http://'+ip_addr+':6666/'
        files = {'media': pixels_data}
```

```python
    try:
        self.root.ids['capture'].text = "Trying to Establish a
        Connection..."
        requests.post(url, files=files)
        self.root.ids['capture'].text = "Capture Again!"
    except requests.exceptions.ConnectionError:
        self.root.ids['capture'].text = "Connection Error! Make Sure
        Server is Active."

def build(self):
    pass

app = PycamApp()
app.run()
```

Figure 3-20 shows the client-side application window.

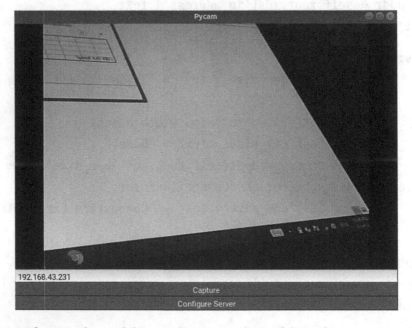

Figure 3-20. *The window of the application after adding the server configuration button*

After preparing the client-side Kivy application, the next step is to prepare the server-side Flask application.

Processing the Received Image at the Server

After successfully building the Kivy application on the client-side, the next step is to prepare the server-side Flask application. It starts by receiving the dimensions of the uploaded images (width and height) and then receives the uploaded images. The application's Python code is shown in Listing 3-30.

There are variables called cam_width and cam_height are defined outside any function. Those variables hold the width and height of the images. The route decorator with the /camSize URL executes the cam_size() function when the button with ID of cam_size in the KV file is pressed. Inside this function, the width and height of the Camera widget are received from the client as arguments using the flask.request.args dictionary. They are assigned to the previously created two variables. In order to use those variables rather than create new ones, we define them as global at the beginning of the function.

Remember to convert the type of the received data to integer before assigning them. The int(float()) variable guarantees that the conversion happens without errors.

Listing 3-30. Restoring Images from the Received Bytes Arrays at the Server

```
import flask
import PIL.Image

app = flask.Flask(import_name="FlaskUpload")

cam_width = 0
cam_height = 0

@app.route('/camSize', methods = ['POST'])
def cam_size():
    global cam_width
    global cam_height

    cam_width = int(float(flask.request.args["width"]))
    cam_height = int(float(flask.request.args["height"]))

    print('Width',cam_width,'& Height',cam_height,'Received Successfully.')

    return "OK"
```

```python
@app.route('/', methods = ['POST'])
def upload_file():
    global cam_width
    global cam_height

    file_to_upload = flask.request.files['media'].read()

    image = PIL.Image.frombytes(mode="RGBA", size=(cam_width, cam_height),
    data=file_to_upload)
    image.save('out.png')

    print('File Uploaded Successfully.')

    return 'SUCCESS'
app.run(host="192.168.43.231", port=6666, debug=True)
```

The upload_file() function is similar to the one used previously in this chapter. It receives the uploaded file using the flask.request.files dictionary. The uploaded file is read using the read() function. The received file is converted to an image using the PIL.Image.frombytes() function. Just for debugging purposes, the image will be saved into a PNG file when developing the client-side application.

After preparing the client and server applications, we can test them according to Figure 3-21. By running the client Kivy application and pressing the button with the ID cam_size, the image size (width and height) will be sent to the server and that button will be removed. After pressing the other button, an image will be captured and sent to the server as a bytes array file. The file is read at the server and returns the bytes array. That array is converted into an image using the PIL.Image.frombytes() function. Figure 3-21 shows that everything is working fine.

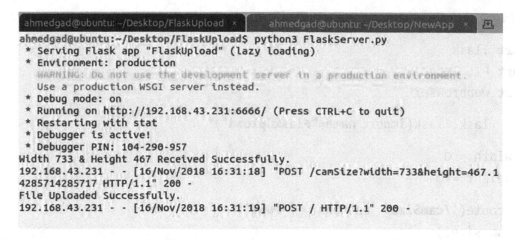

```
ahmedgad@ubuntu: ~/Desktop/FlaskUpload  ×     ahmedgad@ubuntu: ~/Desktop/NewApp  ×
ahmedgad@ubuntu:~/Desktop/FlaskUpload$ python3 FlaskServer.py
 * Serving Flask app "FlaskUpload" (lazy loading)
 * Environment: production
   WARNING: Do not use the development server in a production environment.
   Use a production WSGI server instead.
 * Debug mode: on
 * Running on http://192.168.43.231:6666/ (Press CTRL+C to quit)
 * Restarting with stat
 * Debugger is active!
 * Debugger PIN: 104-290-957
Width 733 & Height 467 Received Successfully.
192.168.43.231 - - [16/Nov/2018 16:31:18] "POST /camSize?width=733&height=467.1
4285714285717 HTTP/1.1" 200 -
File Uploaded Successfully.
192.168.43.231 - - [16/Nov/2018 16:31:19] "POST / HTTP/1.1" 200 -
```

Figure 3-21. *The image dimensions followed by uploading the bytes array are sent to the sever*

At this point, both the Kivy client and the Flask server applications are working well with each other in order to upload a single image. In order to send images continuously from the client to the server, we can build an HTML page that will display the received images.

Saving and Displaying the Received Images Using an HTML Page

Up to this point, we successfully built both the client-side and server-side applications. The client-side application sends the image and its width and height are also sent to the server. The client sends the image as a bytes array to the server. The server uses the received width and height for saving the array as a PNG file.

Because we are interested in displaying the received images, we are going to build a very simple HTML page with an <img> element in which the path of the uploaded image to be displayed is assigned to the src attribute. After receiving and saving the uploaded image as a PNG file, the server application creates the HTML file after updating the src attribute according to the path of the uploaded image. Then, the HTML page is opened in the web browser using the open() function of the webbrowser module. This function accepts the page URL as an argument. The updated server application is shown in Listing 3-31.

Listing 3-31. Displaying the Restored Images on the Server on an HTML Page

```python
import flask
import PIL.Image
import webbrowser

app = flask.Flask(import_name="FlaskUpload")

cam_width = 0
cam_height = 0

@app.route('/camSize', methods = ['POST'])
def cam_size():
    global cam_width
    global cam_height

    cam_width = int(float(flask.request.args["width"]))
    cam_height = int(float(flask.request.args["height"]))

    print('Width',cam_width,'& Height',cam_height,'Received Successfully.')

    return "OK"

@app.route('/', methods = ['POST'])
def upload_file():
    global cam_width
    global cam_height

    file_to_upload = flask.request.files['media'].read()

    image = PIL.Image.frombytes(mode="RGBA", size=(cam_width, cam_height),
    data=file_to_upload)
    image.save('out.png')

    print('File Uploaded Successfully.')

    html_code = '<html><head><title>Displaying Uploaded Image</title></
    head><body><h1>Displaying Uploaded Image</h1><img src="out.png"
    alt="Uploaded Image at the Flask Server"/></body></html>'

    html_url = "/home/ahmedgad/Desktop/FlaskUpload/test.html"
    f = open(html_url,'w')
```

```
f.write(html_code)
f.close()

webbrowser.open(html_url)

return 'SUCCESS'
app.run(host="192.168.43.231", port=6666, debug=True)
```

The HTML code is written as text inside the html_code variable. The formatted code for better visualization is shown in Listing 3-32. In addition to the <img> element, the <h1> element prints a title above it. The HTML code is written into an HTML file according to the specified path in the html_url variable.

Listing 3-32. HTML Page to Display the Images

```
<html>
<head>
<title>Displaying Uploaded Image</title>
</head>
<body>
<h1>Uploaded Image to the Flask Server</h1>
<img src="out.png" alt="Uploaded Image at the Flask Server"/>
</body>
</html>
```

After capturing an image in the client, uploading it to the server, and updating and displaying the HTML page, the result looks as shown in Figure 3-22. Note that the application opens a new tab in the browser for each uploaded image. This will be a trouble later when we try to continuously upload images.

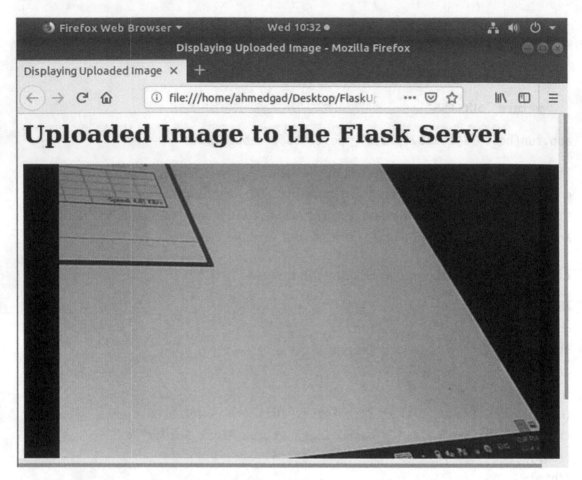

Figure 3-22. *Displaying the uploaded images to the HTML page on the server*

Displaying the Received Image Without Being Saved

In the client-side application, we used the get_region() method in order to avoid saving each uploaded image. We need to do the same thing to the server-side application.

Currently, the server receives the bytes array, saves it to a PNG file using PIL, and displays it in the web browser. We need to remove the in-between step of saving the image as a PNG file. Thus, we need to display the uploaded image as a bytes array directly to the web browser. This is done by inlining the bytes array of the image into the src attribute of the <img> HTML element as a base64 encoded image.

For encoding the bytes array as base64, the base64 Python module is used. Make sure it is installed in your machine. Listing 3-33 shows the updated server-side application.

Note that we are longer need to use PIL. This is because we are not interested in either converting the bytes array into an image or saving the image.

Listing 3-33. Inlining the Bytes Array Into the src Attribute of the <img> HTML Tag

```python
import flask
import webbrowser
import base64

app = flask.Flask(import_name="FlaskUpload")

cam_width = 0
cam_height = 0

@app.route('/camSize', methods = ['POST'])
def cam_size():
    global cam_width
    global cam_height

    cam_width = int(float(flask.request.args["width"]))
    cam_height = int(float(flask.request.args["height"]))

    print('Width',cam_width,'& Height',cam_height,'Received Successfully.')

    return "OK"

@app.route('/', methods = ['POST'])
def upload_file():
    global cam_width
    global cam_height

    file_to_upload = flask.request.files['media'].read()

    print('File Uploaded Successfully.')

    im_base64 = base64.b64encode(file_to_upload)

    html_code = '<html><head><meta http-equiv="refresh" content="1">
<title>Displaying Uploaded Image</title></head><body><h1>Uploaded Image
to the Flask Server</h1><img src="data:;base64,' + im_base64.decode(
'utf8') + '" alt="Uploaded Image at the Flask Server"/></body></html>'
```

```
html_url = "/home/ahmedgad/Desktop/FlaskUpload/test.html"
f = open(html_url,'w')
f.write(html_code)
f.close()

webbrowser.open(html_url)

return 'SUCCESS'
app.run(host="192.168.43.231", port=6666, debug=True)
```

The line used to convert the image to base64 encoding using the b64encode() function is shown next. This function accepts a bytes array and thus it is fed by the uploaded data in the file_to_upload variable.

```
im_base64 = base64.b64encode(file_to_upload)
```

The im_base64 variable holds the base64 encoded image. The value in this variable is assigned to the src attribute of the <img> element as a data URL. The URL used is data:;base64,. Note that the URL does not accept the bytes array directly but after converting it into a string using the encode('utf8') function. You can read more about data URLs.

Remember that we have to convert the uploaded Bytes array image into a PIL image in order to rotate it. Then the PIL image is converted back to a bytes array in order to encode it using base64. By doing this, we do not have to save the image as an external file.

Continuously Uploading Images to the Server

Previously, a single image was uploaded to the server. Now, we want to continuously upload images to the server. To do this, there will be changes to both the client-side and server-side applications.

After clicking the Capture button of the client-side Kivy application, the application enters an infinite while loop. In each iteration, a camera image is captured and sent to the server in a POST HTTP message. The updated Kivy application is shown in Listing 3-34.

Listing 3-34. Client-Side Application for Continuously Capturing and Uploading Images to the Server

```python
import kivy.app
import requests

class PycamApp(kivy.app.App):

    def cam_size(self):
        camera = self.root.ids['camera']
        cam_width_height = {'width': camera.resolution[0], 'height':
        camera.resolution[1]}

        ip_addr = self.root.ids['ip_address'].text
        url = 'http://'+ip_addr+':6666/camSize'

        try:
            self.root.ids['cam_size'].text = "Trying to Establish a
            Connection..."
            requests.post(url, params=cam_width_height)
            self.root.ids['cam_size'].text = "Done."
            self.root.remove_widget(self.root.ids['cam_size'])
        except requests.exceptions.ConnectionError:
            self.root.ids['cam_size'].text = "Connection Error! Make Sure
            Server is Active."

    def capture(self):
        while True:
            camera = self.root.ids['camera']

            pixels_data = camera.texture.get_region(x=camera.x, y=camera.y,
            width=camera.resolution[0], height=camera.resolution[1).pixels

            ip_addr = self.root.ids['ip_address'].text
            url = 'http://'+ip_addr+':6666/'
            files = {'media': pixels_data}
```

```
    try:
            self.root.ids['capture'].text = "Trying to Establish a
            Connection..."
            requests.post(url, files=files)
            self.root.ids['capture].text = "Capture Again!"
        except requests.exceptions.ConnectionError:
            self.root.ids['capture'].text = "Connection Error! Make
            Sure Server is Active."

    def build(self):
        pass

app = PycamApp()
app.run()
```

At the server-side Flask application, a new browser tab is opened for each uploaded image. This is a problem when we want to continuously upload images. To solve this issue, we use a flag variable named html_opened. It is set to False by default, which means no tab is opened. After uploading the first image, it will be set to True and thus the application will not open any other tabs. The updated Flask application is shown in Listing 3-35.

Listing 3-35. Server-Side Application for Continuously Receiving the Uploaded Images and Displaying Them in the Web Browser

```
import flask
import base64
import webbrowser

app = flask.Flask(import_name="FlaskUpload")

cam_width = 0
cam_height = 0

html_opened = False

@app.route('/camSize', methods = ['GET', 'POST'])
def cam_size():
    global cam_width
    global cam_height
```

```
cam_width = int(float(flask.request.args["width"]))
cam_height = int(float(flask.request.args["height"]))

print('Width',cam_width,'& Height',cam_height,'Received Successfully.')

return "OK"

@app.route('/', methods = ['POST'])
def upload_file():
    global cam_width
    global cam_height
    global html_opened

    file_to_upload = flask.request.files['media'].read()

    print('File Uploaded Successfully.')

    im_base64 = base64.b64encode(file_to_upload)

    html_code = '<html><head><meta http-equiv="refresh" content="0.5">
<title>Displaying Uploaded Image</title></head><body><h1>Uploaded
Image to the Flask Server</h1><img src="data:;base64,'+im_base64.
decode('utf8')+'" alt="Uploaded Image at the Flask Server"/></body></
html>'

    html_url = "/home/ahmedgad/Desktop/FlaskUpload/templates/test.html"
    f = open(html_url,'w')
    f.write(html_code)
    f.close()

    if html_opened == False:
        webbrowser.open(html_url)
        html_opened = True

    return "SUCCESS"

app.run(host="192.168.43.231", port=6666, debug=True)
```

Another change in the server application is the use of a <meta> tag to refresh the HTML page every 0.5 seconds.

Controlling Image Upload Rate Using Clock

The previous application used the UI thread to upload images to the server. This hangs the applications and prevents the user from interacting with its widgets.

It is better to do time-consuming operations in another thread rather than the UI thread. In our application, this solution is not feasible. This is because if we created a new thread it will still have to access the Camera widget from the UI thread each time an image is captured.

Another solution is to upload the image to the server in a new thread rather than the UI thread. This makes the UI of the application more responsive than before. Also, we can slow down the process by controlling the rate of uploading images to the server.

Using the kivy.clock.Clock object, we can schedule a function call in the future for execution. Because we are interested in executing the function multiple times in the future, the kivy.clock.Clock.schedule_interval() function is a good option. It accepts the function to be executed and the number of seconds between the two executions. The modified code of the Kivy application is shown in Listing 3-36. The interval is set to 0.5 seconds. Remember to match the number of seconds for uploading an image in the schedule_interval() function and refreshing the HTML page in the <meta> tag.

Listing 3-36. Uploading the Images in a New Thread

```
import kivy.app
import requests
import kivy.clock
import kivy.uix.screenmanager
import threading

class Configure(kivy.uix.screenmanager.Screen):
    pass

class Capture(kivy.uix.screenmanager.Screen):
    pass

class PycamApp(kivy.app.App):
    num_images = 0

    def cam_size(self):
```

```python
        camera = self.root.ids['camera']
        cam_width_height = {'width': camera.resolution[0], 'height':
        camera.resolution[1]}

        ip_addr = self.root.ids['ip_address'].text
        port_number = self.root.ids['port_number'].text
        url = 'http://' + ip_addr + ':' + port_number + '/camSize'

        try:
            self.root.ids['cam_size'].text = "Trying to Establish a
            Connection..."
            requests.post(url, params=cam_width_height)
            self.root.ids['cam_size'].text = "Done."
            self.root.current = "capture"
        except requests.exceptions.ConnectionError:
            self.root.ids['cam_size'].text = "Connection Error! Make Sure
            Server is Active."

    def capture(self):
        kivy.clock.Clock.schedule_interval(self.upload_images, 0.5)

    def upload_images(self, *args):
        self.num_images = self.num_images + 1
        print("Uploading image", self.num_images)

        camera = self.root.ids['camera']

        print("Image Size ", camera.resolution[0], camera.resolution[1])
        print("Image corner ", camera.x, camera.y)

        pixels_data = camera.texture.get_region(x=camera.x, y=camera.y,
        width=camera.resolution[0], height=camera.resolution[1]).pixels

        ip_addr = self.root.ids['ip_address'].text
        port_number = self.root.ids['port_number'].text
        url = 'http://' + ip_addr + ':' + port_number + '/'
        files = {'media': pixels_data}

        t = threading.Thread(target=self.send_files_server, args=(files, url))
        t.start()
```

```
    def build(self):
        pass

    def send_files_server(self, files, url):
        try:
            requests.post(url, files=files)
        except requests.exceptions.ConnectionError:
            self.root.ids['capture'].text = "Connection Error! Make Sure
            Server is Active."

app = PycamApp()
app.run()
```

In this example, a new function called upload_images() is created to hold the
code responsible for capturing and uploading each image. This function increments a
variable named num_images for each uploaded image. Inside the function, just the image
is captured using camera.texture.get_region(). To upload it, a new thread is created
at the end of this function.

Using the Thread class inside the threading module, we can create new threads.
Inside the constructor of that class, the thread target is specified and it can be a
function that is called after the thread is running. If that function accepts arguments, we
can pass them using the args argument of the constructor.

In our application, a callback function named send_files_server() is created and it
accepts the image to be uploaded to the server in addition to the server URL.

After running the Kivy and Flask applications, the messages printed to the terminal
indicate successful execution.

The terminal execution of the Kivy application is shown in Figure 3-23.

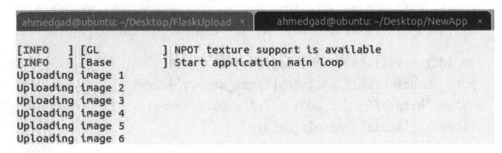

Figure 3-23. *Terminal execution of the client-side application*

Figure 3-24 shows the output of the Flask application.

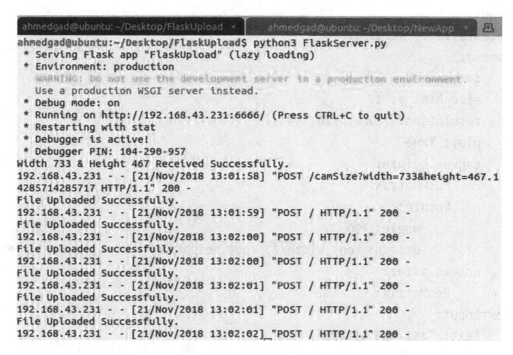

```
ahmedgad@ubuntu: ~/Desktop/FlaskUpload  ×      ahmedgad@ubuntu: ~/Desktop/NewApp  ×
ahmedgad@ubuntu:~/Desktop/FlaskUpload$ python3 FlaskServer.py
 * Serving Flask app "FlaskUpload" (lazy loading)
 * Environment: production
   WARNING: Do not use the development server in a production environment.
   Use a production WSGI server instead.
 * Debug mode: on
 * Running on http://192.168.43.231:6666/ (Press CTRL+C to quit)
 * Restarting with stat
 * Debugger is active!
 * Debugger PIN: 104-290-957
Width 733 & Height 467 Received Successfully.
192.168.43.231 - - [21/Nov/2018 13:01:58] "POST /camSize?width=733&height=467.1
4285714285717 HTTP/1.1" 200 -
File Uploaded Successfully.
192.168.43.231 - - [21/Nov/2018 13:01:59] "POST / HTTP/1.1" 200 -
File Uploaded Successfully.
192.168.43.231 - - [21/Nov/2018 13:02:00] "POST / HTTP/1.1" 200 -
File Uploaded Successfully.
192.168.43.231 - - [21/Nov/2018 13:02:00] "POST / HTTP/1.1" 200 -
File Uploaded Successfully.
192.168.43.231 - - [21/Nov/2018 13:02:01] "POST / HTTP/1.1" 200 -
File Uploaded Successfully.
192.168.43.231 - - [21/Nov/2018 13:02:01] "POST / HTTP/1.1" 200 -
File Uploaded Successfully.
192.168.43.231 - - [21/Nov/2018 13:02:02]_"POST / HTTP/1.1" 200 -
```

Figure 3-24. *Terminal execution of the server-side Flask application*

We have now created a desktop Kivy application that accesses the camera, continuously captures images, uploads them to the Flask server, and displays the captured images in the web browser. The next step is to build the Android application.

Building the Live Camera Preview Android Application

We need to make some changes to the client desktop Kivy application in order to make it suitable as an Android application.

We have to rotate the Camera widget by -90 degrees using the Rotate context instruction because the Android Camera is by default rotated 90 degrees. This was discussed previously in this chapter. The KV file that rotates the widget is shown in Listing 3-37.

Listing 3-37. Rotating the Image 90 Degrees for the Android Application

```
BoxLayout:
    orientation: "vertical"
    Camera:
        id: camera
        size_hint_y: 18
        resolution: (1280, 720)
        play: True
        canvas.before:
            PushMatrix:
            Rotate:
                angle: -90
                origin: root.width/2, root.height/2
        canvas.after:
            PopMatrix:
    TextInput:
        text: "192.168.43.231"
        id: ip_address
        size_hint_y: 1
    Button:
        id: capture
        text: "Capture"
        size_hint_y: 1
        on_press: app.capture()
    Button:
        id: cam_size
        text: "Configure Server"
        size_hint_y: 1
        on_press: app.cam_size()
```

Note that rotating the Camera widget does not mean that the uploaded images to the server are also rotated. This operation just rotates the Camera widget that displays the camera images. The captured images are still rotated by 90 degrees. Because of this, we need to modify the Flask application in order to rotate each captured image by -90 before displaying it in the web browser. This is shown in Listing 3-38. The bytes array is converted into a PIL image, which is rotated by 90 degrees. Finally, the rotated image is converted back into a bytes array in order to be encoded according to base64.

Listing 3-38. Rotating the Captured Images at the Server by 90 Degrees

```python
import flask
import base64
import PIL.Image
import webbrowser

app = flask.Flask(import_name="FlaskUpload")

cam_width = 0
cam_height = 0

html_opened = False

@app.route('/camSize', methods = ['GET', 'POST'])
def cam_size():
    global cam_width
    global cam_height

    cam_width = int(float(flask.request.args["width"]))
    cam_height = int(float(flask.request.args["height"]))

    print('Width',cam_width,'& Height',cam_height,'Received Successfully.')

    return "OK"

@app.route('/', methods = ['POST'])
def upload_file():
    global cam_width
    global cam_height
    global html_opened

    file_to_upload = flask.request.files['media'].read()

    image = PIL.Image.frombytes(mode="RGBA", size=(cam_width, cam_height),
    data=file_to_upload)
    image = image.rotate(-90)
```

```
print('File Uploaded Successfully.')

im_base64 = base64.b64encode(image.tobytes())

html_code = '<html><head><meta http-equiv="refresh" content="0.5">
<title>Displaying Uploaded Image</title></head><body><h1>Uploaded
Image to the Flask Server</h1><img src="data:;base64,'+im_base64.
decode('utf8')+'" alt="Uploaded Image at the Flask Server"/></body>
</html>'

html_url = "/home/ahmedgad/Desktop/FlaskUpload/templates/test.html"
f = open(html_url,'w')
f.write(html_code)
f.close()

if html_opened == False:
    webbrowser.open(html_url)
    html_opened = True

return "SUCCESS"
```

```
app.run(host="192.168.43.231", port=6666, debug=True)
```

After preparing the client-side and server-side applications, we can build the Android application according to the following terminal command.

```
ahmedgad@ubuntu:~/Desktop/NewApp$ buildozer android debug deploy run logcat
```

Make sure to change the path to the application root in which the buildozer.
spec file exists and activate the virtual environment (if you prepared the development environment in a virtual environment).

The window of the Android application is shown in Figure 3-25.

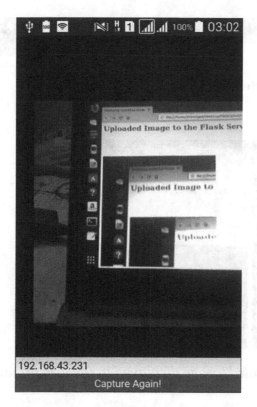

Figure 3-25. *Window of the Android application for continuously uploading images*

After clicking on the Capture button, the Android application captures images continuously and uploads them to the server, where they are displayed in an HTML page. Figure 3-26 shows one of the displayed images.

Figure 3-26. *An uploaded image displayed in the web browser of the server*

Summary

This chapter discussed accessing the Android Camera via the `Camera` widget. Before building the Android application, we created a desktop application to make sure everything worked as expected. We built the Android application using Buildozer. To have permission to access the Android Camera, we had to update the `android.permissions` field inside the `buildozer.init` file . Because the Android Camera is rotated 90 degrees by default, it must be rotated back. This is done using Kivy canvases. Three canvas instances were discussed—`canvas`, `canvas.before`, and `canvas.after`. In order to limit the effect of a given instruction to just certain widgets, the `PushMatrix` and `PopMatrix` instructions were discussed.

After the camera was previewed at the proper angle, images were captured in order to upload them to an HTTP server. The server is created using Flask and runs on a desktop computer. Using the IPv4 address and port number of the server, the `requests` Python library will upload the captured images using the Kivy Android application and an HTTP `POST` message.

At the end of this chapter, we previewed the Android Camera in the server's web browser. The images are saved into the device memory as bytes array, without being saved in the device storage, in order to save time. Such bytes arrays are then uploaded to the server. The server then interprets these bytes array and shows the images via an HTML page in the web browser.

In the next chapter, a more convenient design is created for the live preview project by separating the buttons into different screens. Kivy supports the `Screen` class for building screens and the `ScreenManager` class for managing such screens. We can navigate from one screen to another. To understand how to create an application with multiple screens, the next chapter starts by discussing how to create custom widgets.

## CHAPTER 4

Creating and Managing Multiple Screens

In the previous chapter, we accessed the Android camera using the Camera widget. Kivy canvases were introduced to adjust the camera rotation. In order to limit the effect of a given canvas instruction to just some widgets, the PushMatrix and PopMatrix instructions were discussed. After that, we created an Android Kivy application to continuously capture images and send them to the Flask server, which displays them in an HTML page.

In this chapter, we create a more convenient design by separating the buttons into different screens. Kivy supports the Screen class for building screens and the ScreenManager class for managing such screens. We can navigate from one screen to another. The chapter starts by discussing how to create custom widgets, which will help us understand how to create an application with multiple screens.

Modifying Existing Widgets

Kivy supports a number of existing widgets, such as Button, Label, TextInput, and more. It supports modifying existing widgets to override their default behaviors. We can use the Label widget as a test case.

The Label class includes some default values as its properties. For example, the text property is by default set to an empty string, the text color is white, and default font size is equal to 15 SP (Scale-Independent Pixel). We will override these three properties according to the KV code shown in Listing 4-1. The text of the label is set to "Hello", the text color is red, and the font size is 50 SP.

© Ahmed Fawzy Mohamed Gad 2019
A. F. M. Gad, *Building Android Apps in Python Using Kivy with Android Studio,*
https://doi.org/10.1007/978-1-4842-5031-0_4

Listing 4-1. Overriding Properties of a Widget Inside the KV File

```
Label:
    text: "Hello"
    color: 1,0,0,1
    font_size: "50sp"
```

The Python code shown in Listing 4-2 creates a new class named `TestApp` that extends the `kivy.app.App` class for building a new application. It assumes you saved the previous KV code in a file named `test.kv`.

Listing 4-2. The Generic Code for Building a Kivy Application

```
import kivy.app

class TestApp(kivy.app.App):
    pass

app = TestApp()
app.run()
```

When you run the application, you'll see the result in Figure 4-1. The properties are changed correctly. You might wonder if these properties stay changed for new labels. We can answer this question by creating a new label widget.

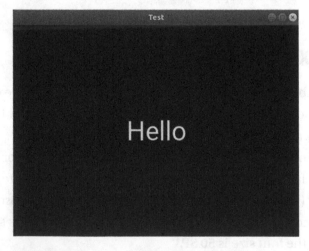

Figure 4-1. *A Kivy application with only a Label widget*

The new KV code in Listing 4-3 creates a BoxLayout that holds both labels. The first label has its properties set according to the previous example, while the second label has just its text changed to "Second Label".

Listing 4-3. Adding Two Label Widgets to the Application Inside the BoxLayout Root Widget

```
BoxLayout:
    Label:
        text: "Hello"
        color: 1,0,0,1
        font_size: "50sp"
    Label:
        text: "Second Label"
```

After running the application, the second label does not have the color and font size changed according to the window in Figure 4-2. The reason is that both labels are independent instances of the Label class. When a new instance is created, it inherits the default values of the properties from the Label class. If some properties are changed for a given instance, it does not mean they will be changed for the other instances. In order to make both labels have the same text color, we can change the color property of the Label class. As a result, all of its instances will inherit this color.

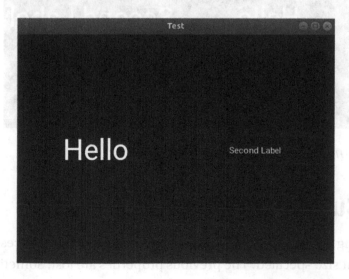

Figure 4-2. *Overriding the properties of only one Label widget while leaving the other set to the default*

In order to edit a class in the KV file, the class name is inserted between <> without any indentation. The KV file in Listing 4-4 overrides the text color and font size properties of the Label class. By creating two instances of the Label class, both will inherit the text color and font size according to Figure 4-3.

Listing 4-4. Editing a Class in the KV Language

```
BoxLayout:
    Label:
        text: "Hello"
    Label:
        text: "Second Label"

<Label>:
    color: 1,0,0,1
    font_size: "50sp"
```

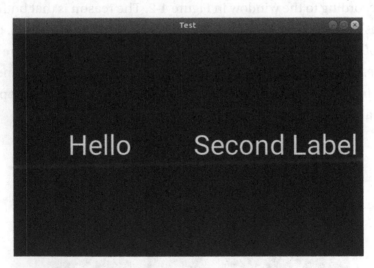

Figure 4-3. *Changing the properties of all Label widgets*

Creating Custom Widgets

The code in Listing 4-4 modifies the Label class so that all of its instances will have the text color and font size specified. The previous properties are lost. Sometimes, we might be interested in the previous default values for such properties.

In order to keep the previous properties of the Label class, we can create a new custom class that extends the Label class. This custom class inherits the default properties of the parent Label class and we can also modify some of its properties.

The KV code in Listing 4-5 creates a new custom class named CustomLabel that inherits the Label class. As a result, if you need to create a label widget with the default properties, you instantiate the Label class. To use the modified properties, instantiate the CustomLabel class. In this example, the first label is an instance of the CustomLabel class in which the text color and font size are changed. The second label is an instance of the Label class with the default values for these two properties.

Listing 4-5. Creating a New Custom Label Widget by Extending the Label Class Inside the KV File

```
BoxLayout:
    CustomLabel:
        text: "Hello"
    Label:
        text: "Second Label"
<CustomLabel@Label>:
    color: 1,0,0,1
    font_size: "50sp"
```

The result after running the application with this KV file is shown in Figure 4-4.

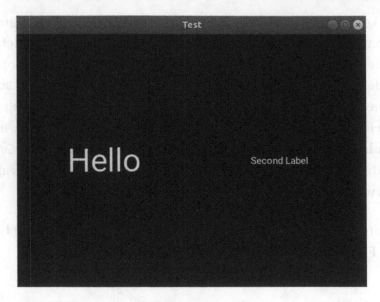

Figure 4-4. *Using the custom Label widget*

Defining Custom Classes in Python

In Listing 4-5, a new custom class named CustomLabel is created in the KV file that both inherits the Label class and modifies some of its properties. Doing the inheritance inside the KV file limits the capabilities of the new class, as we can't write functions inside it.

We can create the new class and do the inheritance inside the Python code. Then, we will just refer to this class in the KV file to modify its properties. This helps when writing Python functions inside the new custom class. The example in Listing 4-6 creates a new empty class named CustomLabel that extends the Label class.

Listing 4-6. Inheriting the Label Class Within the Python File

```python
import kivy.app
import kivy.uix.label

class CustomLabel(kivy.uix.label.Label):
    pass

class TestApp(kivy.app.App):
    pass

app = TestApp()
app.run()
```

The content of the `test.kv` file is shown in Listing 4-7. Note that we simply referred to the existing class inside KV rather than creating it, as in the previous section.

Listing 4-7. Referring to the Custom Class Created in the Python File Inside the KV File

```
BoxLayout:
    CustomLabel:
        text: "Hello"
    Label:
        text: "Second Label"

<CustomLabel>:
    color: 1,0,0,1
    font_size: "50sp"
```

We can change the previous application a bit by creating a class in the Python file named MyLayout that extends the BoxLayout class, as shown in Listing 4-8. Because this class inherits the BoxLayout class, we can use it everywhere that BoxLayout is used. For example, we can replace the BoxLayout inside the KV file with the new class.

Listing 4-8. Creating a New Custom Layout by Extending the BoxLayout Class

```
import kivy.app
import kivy.uix.label
import kivy.uix.boxlayout

class CustomLabel(kivy.uix.label.Label):
    pass

class MyLayout(kivy.uix.boxlayout.BoxLayout):
    pass

class TestApp(kivy.app.App):
    def build(self):
        return MyLayout()

app = TestApp()
app.run()
```

The new KV file is given in Listing 4-9. It references the custom `MyLayout` class by adding its name between `<>`. This class has two child widgets, which are `CustomLabel` and `Label`.

Note that we have to define the `build()` function inside the `TestApp` class to return an instance of the `MyLayout` class. This is because the KV file does not return a layout by itself for the `TestApp`. The KV file simply creates two custom widgets named `MyLayout` and `CustomLabel`.

Listing 4-9. Referencing the Custom BoxLayout Class Inside the KV File

```
<MyLayout>:
    CustomLabel:
        text: "Hello"
    Label:
        text: "Second Label"

<CustomLabel>:
    color: 1,0,0,1
    font_size: "50sp"
```

We can also return the layout for the TestApp class inside the KV file according to the KV file in Listing 4-10. In this case, the KV file defined two new widgets and returned a widget named `MyLayout`. This widget represents the layout of the `TestApp` class. The Python code does not have to implement the `build()` function at the current time.

Listing 4-10. Using the Custom BoxLayout Class

```
MyLayout:

<MyLayout>:
    CustomLabel:
        text: "Hello"
    Label:
        text: "Second Label"

<CustomLabel>:
    color: 1,0,0,1
    font_size: "50sp"
```

At this point, we are able to create a new class inside the Python file that extends a widget class, refer to it inside the KV file, and modify some of its properties. This enables us to start learning how to create an application with multiple screens.

Creating and Managing Screens

Previously, a custom class is created that extends the `kivy.app.App` class when building an application. The application has a window in which we can add widgets. All widgets are inside a single screen. Sometimes, we need to organize the widgets of the same application into different screens where each screen does a different job. The screen in Kivy is similar to the activity in Android. An Android application can have more than one activity and a Kivy application might have more than one screen.

In order to create a screen, rather than extending the `kivy.app.App` class, we will extend the `kivy.uix.screenmanager.Screen` class. Listing 4-11 shows the Python file that creates two classes named `Screen1` and `Screen2`, one for each screen, extending the Screen class. There is also an application class named `TestApp`.

Listing 4-11. Creating Two Screens by Extending the Screen Class

```
import kivy.app
import kivy.uix.screenmanager

class Screen1(kivy.uix.screenmanager.Screen):
    pass

class Screen2(kivy.uix.screenmanager.Screen):
    pass

class TestApp(kivy.app.App):
    pass

app = TestApp()
app.run()
```

From the Python code in Listing 4-11, two empty screens are created. Their layout is given in the `test.kv` file associated with this application, as shown in Listing 4-12. Note that the screen class name is written between `<>`. Each screen has a name property. The names of the two screens are `Screen1` and `Screen2`. There is a screen manager that has two children, which are the two screens. The screen manager has a property named `current`, which tells which screen is currently active in the window. This property accepts the screen name. Each screen has a property named `manager`, which corresponds to the manager of the screen. We can use it to access the manager within the KV file.

Listing 4-12. Defining the Layout of the Two Screens and Adding Them as Children to the ScreenManager Class

```
ScreenManager:
    Screen1:
    Screen2:

<Screen1>:
    name: "Screen1"
    Button:
        text: "Button @ Screen 1"
        on_press: root.manager.current = "Screen2"

<Screen2>:
    name: "Screen2"
    Button:
        text: "Button @ Screen 2"
        on_press: root.manager.current = "Screen1"
```

In order to move from one screen to another, we add a button to each screen. When such a button is pressed, the current screen is changed using the `root.manager.current` property. Inside the first screen, the current screen changes to the second screen. The reverse occurs to the second screen. If the current property is not specified inside the screen manager, it defaults to the first screen inside the manager. Figure 4-5 shows the result after running the application.

Figure 4-5. *The first screen added within the ScreenManager appears as the application startup screen*

Clicking on the button changes the current screen using the current property of the manager, as shown in Figure 4-6.

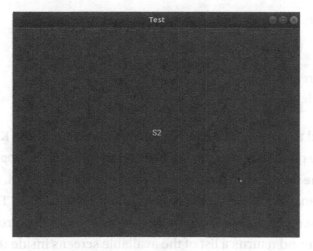

Figure 4-6. *Moving from one screen to another*

We can explicitly specify which screen should be displayed when the application starts using the current property, as given in Listing 4-13. When the application starts, it will open the second screen.

Listing 4-13. Using the current Property of the ScreenManager Class to Explicitly Specify the Startup Screen

```
ScreenManager:
    current: "Screen2"
    Screen1:
    Screen2:
```

Accessing Widgets Inside Screens

After adding the screens and their manager, the widget tree appears as follows. The root widget is the ScreenManager, which holds two child Screen widgets. Each screen has a Button widget. It is important to study the widget tree in order to understand how to access a specific widget in the tree.

- Application
  - Root (ScreenManager)
    - Screen1
      1. Button
    - Screen2
      1. Button

Assume we need to access the button in the first screen from the KV file. How do we do that? First, we need to access the application itself using the app keyword. Then, the root widget of the application is accessed using the root keyword. Note that the root widget is a ScreenManager. Thus, the current command is app.root. The children inside the root widget are screens that can be accessed using the screens property. The app.root.screens command returns a list of the available screens inside the **manager,** as shown in the next line:

```
[<Screen name='Screen1'>, <Screen name='Screen2'>]
```

The first screen is the first element of the list and thus can be accessed using index 0. Thus, the complete command for accessing the first screen is app.root.screens[0].

After accessing the target screen, we can access the button inside it using the `ids` dictionary as we used to do. Assume that the button has an ID of `b1`. If that's the case, the command to access the button would be as follows:

```
app.root.screens[0].ids["b1"]
```

After creating screens and controlling them using the screen manager, we can start modifying the previous project to separate the widgets across two screens.

Modifying the Live Camera Capture Application to Use Screens

In Listings 3-37 and 3-38 of the previous chapter, a Kivy application was created that continuously captures images to be sent to an HTTP server where the received images are displayed in an HTML page. All widgets required to configure and capture images were on the same screen. In this section, they will be separated into different screens, where each screen has a specific job to do.

The first step is to prepare the Python file by adding two screens. The first screen configures the server with the width and height of the images to be captured. The second screen captures the images and sends them to the server. The modified Python code in Listing 4-14 has two new classes, named `Configure` and `Capture`, that extend the `Screen` class.

Listing 4-14. Using the Screen Class to Redesign the Live Camera Capture Application Created in Listing 3-37

```python
import kivy.app
import requests
import kivy.clock
import kivy.uix.screenmanager
import threading

class Configure(kivy.uix.screenmanager.Screen):
    pass

class Capture(kivy.uix.screenmanager.Screen):
    pass
```

```python
class PycamApp(kivy.app.App):
    num_images = 0

    def cam_size(self):
        camera = self.root.screens[1].ids['camera']
        cam_width_height = {'width': camera.resolution[0], 'height':
        camera.resolution[1]}

        ip_addr = self.root.screens[0].ids['ip_address'].text
        port_number = self.root.screens[0].ids['port_number'].text
        url = 'http://' + ip_addr + ':' + port_number + '/camSize'

        try:
            self.root.screens[0].ids['cam_size'].text = "Trying to
            Establish a Connection..."
            requests.post(url, params=cam_width_height)
            self.root.screens[0].ids['cam_size'].text = "Done."
            self.root.current = "capture"
        except requests.exceptions.ConnectionError:
            self.root.screens[0].ids['cam_size'].text = "Connection Error!
            Make Sure Server is Active."

    def capture(self):
        kivy.clock.Clock.schedule_interval(self.upload_images, 1.0)

    def upload_images(self, *args):
        self.num_images = self.num_images + 1
        print("Uploading image", self.num_images)

        camera = self.root.screens[1].ids['camera']

        print("Image Size ", camera.resolution[0], camera.resolution[1])
        print("Image corner ", camera.x, camera.y)

        pixels_data = camera.texture.get_region(x=camera.x, y=camera.y,
        width=camera.resolution[0], height=camera.resolution[1]).pixels

        ip_addr = self.root.screens[0].ids['ip_address'].text
        port_number = self.root.screens[0].ids['port_number'].text
```

```
        url = 'http://' + ip_addr + ':' + port_number + '/'
        files = {'media': pixels_data}

        t = threading.Thread(target=self.send_files_server, args=(files, url))
        t.start()

    def build(self):
        pass

    def send_files_server(self, files, url):
        try:
            requests.post(url, files=files)
        except requests.exceptions.ConnectionError:
            self.root.screens[1].ids['capture'].text = "Connection Error!
            Make Sure Server is Active."

app = PycamApp()
app.run()
```

The widget tree inside the KV file is shown here. Note that the widgets are separated across the two screens.

- Application
  - Root (ScreenManager)
  - Configure Screen
    - BoxLayout
      - Label
      - TextInput (ip_address)
      - TextInput (port_number)
      - Button (cam_size)
  - Capture Screen
    - BoxLayout
      - Camera (camera)
      - Button (capture)

The KV file of the application is shown in Listing 4-15, where each screen has a BoxLayout for grouping its widgets. The Configure screen has a Label widget that displays instructions for the users. There are two TextInput widgets where the user enters the IPv4 address and port number at which the server listens for requests. It also includes the Button widget used for sending a POST message by the dimensions of the camera. The Capture screen includes the Camera widget itself and a button for starting capturing images.

Both screens are grouped under the ScreenManager. Note that the Configure screen is the first screen to be added to the manager and thus it will be displayed when the application starts.

Listing 4-15. The KV File of the Live Camera Capture Project After Using Screens

```
ScreenManager:
    Configure:
    Capture:

<Capture>:
    name: "capture"
    BoxLayout:
        orientation: "vertical"
        Camera:
            id: camera
            size_hint_y: 18
            resolution: (1024, 1024)
            allow_stretch: True
            play: True
            canvas.before:
                PushMatrix:
                Rotate:
                    angle: -90
                    origin: root.width/2, root.height/2
            canvas.after:
                PopMatrix:
```

```
        Button:
            id: capture
            font_size: 30
            text: "Capture"
            size_hint_y: 1
            on_press: app.capture()

<Configure>:
    name: "configure"
    BoxLayout:
        orientation: "vertical"
        Label:
            text: "1) Enter the IPv4 address of the server.\n2) Enter the
            port number. \n3) Press the Configure Server button. \nMake
            sure that the server is active."
            font_size: 20
            text_size: self.width, None
            size_hint_y: 1
        TextInput:
            text: "192.168.43.231"
            font_size: 30
            id: ip_address
            size_hint_y: 1
        TextInput:
            text: "6666"
            font_size: 30
            id: port_number
            size_hint_y: 1
        Button:
            id: cam_size
            font_size: 30
            text_size: self.width, None
            text: "Configure Server"
            size_hint_y: 1
            on_press: app.cam_size()
```

Once the user presses the button that configures the server, the Camera widget dimensions are returned and a POST message is sent to the server based on the IPv4 address and port number retrieved from the TextInput widgets. The first screen is shown in Figure 4-7.

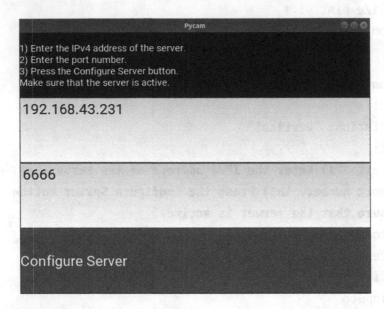

Figure 4-7. *The main screen of the application where the IP and port number are specified*

After the message is sent successfully, the current screen of the manager is changed to the Capture screen, which is shown in Figure 4-8. In this screen, the user can press the capture button in order to start capturing and sending the captured images to the server.

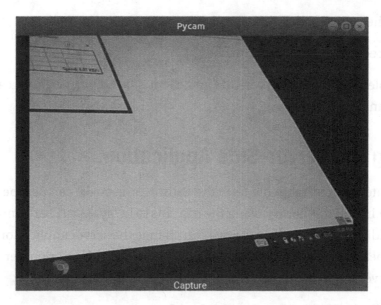

Figure 4-8. *The second screen in the application where images can be captured and sent to the server*

Notice how the widgets are accessed from the widget tree. As discussed in the previous section, the ScreenManager is the root and has two screens. Each screen has a number of widgets that can be accessed using their IDs. For example, the Camera widget can be accessed from the KV file using the following command.

```
app.root.screens[1].ids['camera']
```

In this project, we are not interested in referring to widgets from the KV file but from the Python file. For example, the Camera widget must be accessed from inside the cam_size() function of the PycamApp class. In this case, the difference compared to the previous command is how to access the application. It can be referenced using the self keyword. Thus, the command used to access the Camera widget inside Python is as follows.

```
self.root.screens[1].ids['camera']
```

We use screen with index 1 because the Camera widget resides inside it. This way, we successfully access a widget from the second screen with index 1. If we need to access the TextInput widget with an ID of ip_address, which is available in the first screen from the Python code, then the next command is used. Just specify the index of the screen in addition to the ID of the widget.

```
self.root.screens[0].ids['ip_address']
```

To access the port number, the next command is used:

```
self.root.screens[0].ids['port_number']
```

After completing both the server-side and client-side applications, we can start publishing them.

Publishing the Server-Side Application

In order to create an executable file from the Python project, we can use the PyInstaller library. We can install this library using the `pip install pyinstaller` command.

Before building the executable file, we can change the server application a bit. This is because it does not allow us to change the IPv4 address and port number. We used to execute the server application using the following terminal command:

```
ahmedgad@ubuntu:~/Desktop$ python3 FlaskServer.py
```

When executing a Python file from the terminal, some arguments are passed to it inside `sys.argv` list. If no arguments are specified in the terminal, then there will be a single item in the list holding the Python script name, which can be accessed according to the following command:

```
sys.argv[0]
```

Arguments can be listed after the name of the Python script. For example, the next command passes the IPv4 address and port number as arguments to the script.

```
ahmedgad@ubuntu:~/Desktop$ python3 FlaskServer.py 192.168.43.231 6666
```

In order to access the IPv4 address inside the Python script and store it in a variable named `ip_address`, we use the next command. The index 1 is used, as it is the second argument in the list.

```
ip_address = sys.argv[1]
```

By the same token, the port number is stored into the `port_number` variable using the next command. Note that index 2 is used.

```
port_number = sys.argv[2]
```

The new Python code of the server application listed in Listing 4-16 fetches the IPv4 address and port number from the terminal arguments. Inside the `app.run()` method, the host and port arguments take their values from the `ip_address` and `port_number` variables rather than being statically defined.

Listing 4-16. Modified Python Code for the Server-Side Application for Fetching the IPv4 Address and Port Number from the Command-Line Arguments

```python
import flask
import PIL.Image
import base64
import webbrowser
import sys
import os

app = flask.Flask(import_name="FlaskUpload")

cam_width = 0
cam_height = 0

html_opened = False

@app.route('/camSize', methods = ['GET', 'POST'])
def cam_size():
    global cam_width
    global cam_height

    cam_width = int(float(flask.request.args["width"]))
    cam_height = int(float(flask.request.args["height"]))

    print('Width',cam_width,'& Height',cam_height,'Received Successfully.')

    return "OK"

@app.route('/', methods = ['POST'])
def upload_file():
    global cam_width
    global cam_height
    global html_opened
```

121

```python
file_to_upload = flask.request.files['media'].read()

image = PIL.Image.frombytes(mode="RGBA", size=(cam_width, cam_height),
data=file_to_upload)
image = image.rotate(-90)
print('File Uploaded Successfully.')

im_base64 = base64.b64encode(image.tobytes())

html_code = '<html><head><meta http-equiv="refresh"
content="1"><title>Displaying Uploaded Image</title></
head><body><h1>Uploaded Image to the Flask Server</h1><img
src="data:;base64,'+im_base64.decode('utf8')+'" alt="Uploaded Image at
the Flask Server"/></body></html>'

# The HTML page is not required to be opened from the Python code but
open it yourself externally.
html_url = os.getcwd()+"/templates/test.html"
f = open(html_url,'w')
f.write(html_code)
f.close()

if html_opened == False:
    webbrowser.open(html_url)
    html_opened = True

return "SUCCESS"
ip_address = sys.argv[1]#"192.168.43.231"
port_number = sys.argv[2]#6666
app.run(host=ip_address, port=port_number, debug=True, threaded=True)
```

After being installed, the project can be turned into an executable file using the following command. Just replace the <python-file-name> with the Python filename of the server. The --onefile option makes PyInstaller generate a single binary file. Be sure to execute this command at the location in which the Python file is executed as long as its complete path is not specified.

```
pyinstaller --onefile <python-file-name>.py
```

After the command completes, the binary file will exist inside the dist folder, named according to the Python filename. PyInstaller creates an executable file for the OS being used. If the command is executed in a Linux machine, then a Linux binary is produced. If executed in Windows, then a Windows executable file (.exe) will be created.

The executable file can be hosted in a repository of your choice, where users can download and run the server. The Linux executable file is available at this page for download under the the CamShare name: https://www.linux-apps.com/p/1279651. Thus, in order to run the server, just download the file and run the following terminal command. Remember to change the path of the terminal according to the current path of CamShare.

```
ahmedgad@ubuntu:~/Desktop$ python3 CamShare 192.168.43.231 6666
```

Publishing the Client-Side Android Application to Google Play

The previous APK files are just for debugging and cannot be published at Google Play, as it accepts only release APKs. In order to create a release version of the application, we use this next command:

```
ahmedgad@ubuntu:~/Desktop$ buildozer android release
```

It is important to sign your release APK in order for it to be accepted at Google Play. For the instructions about signing an APK, read this page: https://github.com/kivy/kivy/wiki/Creating-a-Release-APK. Also remember to set the target API level to at least 26, as discussed previously.

You can create a developer account at Google Play for publishing your own applications. The CamShare Android application is available here: https://play.google.com/store/apps/details?id=camshare.camshare.myapp.

You can download the Android app, connect it to the server, and capture images that will be displayed in an HTML page on the server.

Summary

As a summary, this chapter introduced building custom widgets by extending the Kivy widgets. This allows us to edit their properties once and use them many times. The chapter also introduced the Screen and ScreenManager classes for organizing the application widgets across multiple screens. For specifying which screen appears as soon as the application starts, the current property of the ScreenManager is set to the name of the desired screen. The screens are used to redesign the interface of the live camera capture project in Chapter 3.

In the next chapter, the Kivy concepts introduced in this and all previous chapters will be applied in order to create a multi-level cross-platform game in which a player collects a number of randomly distributed coins across the screen. There will be monsters that try to kill the player. The next chapter moves the reader from zero to hero by making game development very easy and explaining each line of code.

## CHAPTER 5

Building Your First Multi-Level Game in Kivy

The previous chapter introduced Kivy so that we could start building cross-platform applications. As a way of applying concepts, we created an Android application that captures images and continuously sends them to a Flask server.

This chapter applies those same concepts by creating a multi-level cross-platform game where the player has a mission in each level, which is to collect coins that are randomly distributed on the screen. Monsters try to kill the players while they collect coins. The game functions successfully in different platforms without us having to change a single line of code. Before you learn how to build the game, we'll introduce some new concepts, including FloatLayout and animation.

FloatLayout

In the previous chapters, the BoxLayout widget was used to group multiple widgets. Widgets are added to this layout in an ordered manner—either horizontally or vertically according to the orientation. The widget's size is calculated by the layout with minor control over it. In the game we are going to create in this chapter, some widgets will not follow a pre-defined order. We need to customize their sizes and freely move them to any position. For example, the main character is placed according to the touch position. For this reason, we will use the FloatLayout widget. It places the widgets according to the x and y coordinates specified in each widget.

Listing 5-1 shows the generic code used to build a Kivy application in which the child class is named TestApp.

© Ahmed Fawzy Mohamed Gad 2019
A. F. M. Gad, *Building Android Apps in Python Using Kivy with Android Studio*,
https://doi.org/10.1007/978-1-4842-5031-0_5

Listing 5-1. Generic Python Code to Build a Kivy Application

```python
import kivy.app

class TestApp(kivy.app.App):
    pass

app = TestApp()
app.run()
```

Based on the class name, the KV file must be named `test.kv` in order to implicitly detect it. The `test.kv` file content is shown in Listing 5-2. There is just a `FloatLayout` widget with a child, called `Button`. Note that there are two important fields in the `Button` widget—`size_hint` and `pos_hint`. Compared to `BoxLayout`, widgets added using `FloatLayout` may not extend the entire width or height of the screen.

Listing 5-2. KV File with FloatLayout as the Root Widget

```
FloatLayout:
    Button:
        size_hint: (1/4, 1/4)
        pos_hint: {'x': 0.5,'y': 0.5}
        text: "Hello"
```

If you run the application, you'll see the window in Figure 5-1.

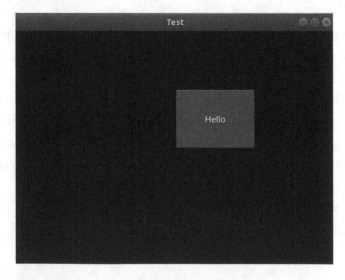

Figure 5-1. *A button added within the FloatLayout*

By default, widgets are added to the parent in the (0, 0) position, which corresponds to the bottom-left corner of the window. Thus, we need to move the widgets to avoid placing them on top of each other. The pos_hint field accepts a dictionary with two fields specifying the distance between the bottom-left corner of a widget and the bottom-left corner of the window. The distance is relative to the parent size.

A value of 0.5 for x means that the button will be away from the left size of the window by 50% of the parent width. A value of 0.5 for y means that the button will be far from the bottom of the window by 50% of the parent height. This way, the bottom-left corner of the Button widget starts at the center of the layout. Note that relative positioning is an efficient way of working with screens of differing sizes.

The size_hint field specifies the size of the widget relative to its parent size. It accepts a tuple that holds the relative width and height of the widget. In this example, the width and height of the button are set to 1/4, meaning that the button size is 40% of the parent size (i.e., one quarter).

Note that the pos_hint and size_hint fields are not guaranteed to change the size or the position of the widget. A widget just gives a hint to the parent that it prefers to set its position and size according to the values specified. Some layouts listen to its request as to which layouts neglect it. In the previous example, if FloatLayout is replaced with BoxLayout according to the code in Listing 5-3, some hints are not applied to the layout according to Figure 5-2. Note that the default orientation is horizontal.

Listing 5-3. Horizontal BoxLayout Orientation Does Not Listen to the Width Hint

```
BoxLayout:
    Button:
        size_hint: (1/4, 1/4)
        pos_hint: {'x': 0.5,'y': 0.5}
        text: "Hello"
```

Because the button is the only child in its horizontal BoxLayout parent, its bottom-left corner is expected to start from the (0, 0) position. According to Figure 5-2, the button does not start from the (0, 0) position. Its x coordinate starts at 0 as expected, but its y coordinate starts at half the parent's height. As a result, the parent just listened to the hint about the Y position.

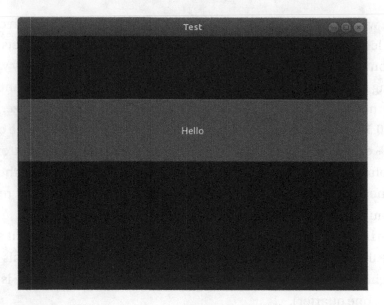

Figure 5-2. *The button extends the entire width of the screen even if the width hint is set to 1/4*

Regarding the button size, it is expected to cover the entire window, as it is the only child in the parent. This does not happen in the previous example. The height is 1/4 of the parent height, but the width extends the entire width of the parent.

As a summary of what happened when the pos_hint and size_hint fields are used with BoxLayout, just the height and the Y position change, while the width and the X position did not change. The reason is that a BoxLayout with a horizontal orientation just listens to hints related to the Y axis (e.g., height and Y position). If vertical orientation is used according to Listing 5-4, the width and the X position will change, but the height and the Y position will not according to Figure 5-3. This is why FloatLayout is used to dynamically position and size widgets.

Listing 5-4. Vertical Orientation for BoxLayout Does Not Listen to the Height Hint

```
BoxLayout:
    orientation: 'vertical'
    Button:
        size_hint: (1/4, 1/4)
        pos_hint: {'x': 0.5,'y': 0.5}
        text: "Hello"
```

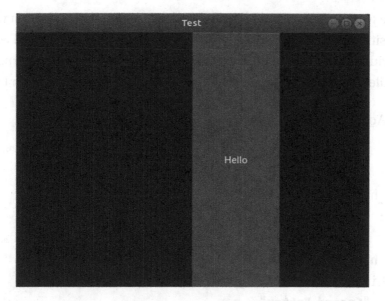

Figure 5-3. *The button extends the entire height of the screen even if the height hint is set to 1/4*

Note that the `pos_hint` field changes both the X and Y coordinates. If we are just interested in changing one rather than both, we can specify that in the dictionary. Note that there are other items to be specified in the dictionary such as `top`, `right`, `center_x`, and `center_y`.

In addition, the `size_hint` field specifies both the width and height. We can use `size_hint_x` to specify the width or the `size_hint_y` for the height. Because `BoxLayout` with a horizontal orientation does not change the X position and the width of the widgets, we can avoid specifying them. Listing 5-5 produces the same results while using fewer hints.

Listing 5-5. Just Specifying the Height Hint Using size_hint_y

```
BoxLayout:
    Button:
        size_hint_y: 1/4
        pos_hint: {'y': 0.5}
        text: "Hello"
```

Assume we want to add two widgets to the FloatLayout, where the first widget starts at the (0, 0) position and extends to the center of the layout and the second widget starts at 75% of the width and height of the parent and extends to its top-right corner. Listing 5-6 shows the KV file needed to build such a widget tree. The result is shown in Figure 5-4.

Listing 5-6. Adding Two Buttons Inside FloatLayout

```
FloatLayout:
    Button:
        size_hint: (0.5, 0.5)
        text: "First Button"
    Button:
        size_hint: (0.25, 0.25)
        pos_hint: {'x': 0.75, 'y': 0.75}
        text: "Second Button"
```

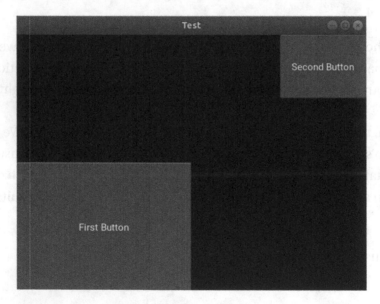

Figure 5-4. *Adding two buttons within FloatLayout*

The size_hint field of the first button size is set to 0.5 to both the width and the height to make its size 50% of the window size. Its pos_hint is omitted, as the widget by default starts at the (0, 0) position.

The pos_hint of the second button is set to 0.75 for both x and y to make it start at a location that's 75% of the way from the width and height of the parent. Its size_hint is set to 0.25 to make the button extend to the top-right corner.

Animation

In order to create a game in Kivy, animation is essential. It makes things go smoothly. For example, we might be interested in animating a monster that moves along a specified path. In Kivy, animations can be created by simply using the kivy.animation.Animation class. Let's create an application with an image widget and animate it by changing its position.

Listing 5-7 shows the KV file of the application. The root widget is FloatLayout, which contains two child widgets. The first child widget is an Image widget with the character_image ID and it displays the image specified by the source field. When set to True, the allow_stretch property stretches the image to cover the entire area of the Image widget.

There is a Button widget that starts the animation when it's pressed. For this reason, a function named start_char_animation() within the Python file is associated with the on_press event.

Listing 5-7. Adding an Image to the Widget Tree Using the Image Widget

```
FloatLayout:
    Image:
        id: character_image
        size_hint: (0.15, 0.15)
        pos_hint: {'x': 0.2, 'y': 0.6}
        allow_stretch: True
        source: "character.png"
    Button:
        size_hint: (0.3, 0.3)
        text: "Start Animation"
        on_press: app.start_char_animation()
```

The Python file implementation is shown in Listing 5-8. Inside the TestApp class, the start_char_animation() function is implemented. An instance of the kivy.animation. Animation class is created in the char_animation variable. This class accepts the properties of the target widget to be animated. Because we are interested in changing the position of the Image widget, the pos_hint property is given as an input argument to the Animation class constructor.

Note that it is not possible to animate a property that's not defined in the widget. For example, we cannot animate the width property because it is not defined in the widget.

Listing 5-8. Building and Starting the Animation Over an Image

```python
import kivy.app
import kivy.animation

class TestApp(kivy.app.App):

    def start_char_animation(self):
        character_image = self.root.ids['character_image']
        char_animation = kivy.animation.Animation(pos_hint={'x':0.8,
        'y':0.6})
        char_animation.start(character_image)

app = TestApp()
app.run()
```

In order to animate a property in a widget, we must provide the property name and its new value. The animation starts from the previous value of the property, which in this case is the value specified by the pos_hint field in the KV file, which is {'x': 0.2, 'y': 0.6}, and ends at the value specified in the constructor of the Animation class, which is {'x': 0.8, 'y': 0.6}. Because there is just a change in the x position, the image will move horizontally.

The start() method of the Animation class is called in order to start the animation. This method accepts the ID of the target widget in which we are looking to animate the properties specified in the Animation class constructor.

When we click the Button widget, the start_char_animation() function is executed and the animation starts. Figure 5-5 shows how the window appears, before and after pressing the button. By default, the animation takes one second to complete. This time can be changed using the duration argument.

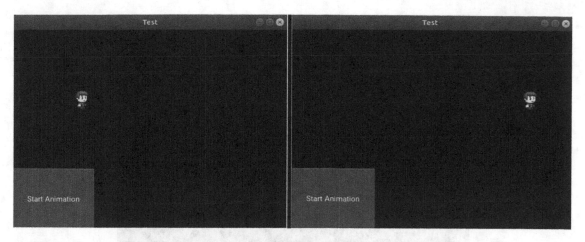

Figure 5-5. *Before and after the animation starts. The left figure shows the original image and the right figure shows the result after the animation ends*

Note that we can animate multiple properties at the same animation instance. This is done by separating the different properties with commas. Listing 5-9 animates both the size and position of the image. The size is doubled by changing it from (0.15, 0.15) to (0.2, 0.2).

Listing 5-9. Animating Multiple Properties Within the Same Animation Instance

```python
import kivy.app
import kivy.animation

class TestApp(kivy.app.App):

    def start_char_animation(self):
        character_image = self.root.ids['character_image']
        char_animation = kivy.animation.Animation(pos_hint={'x': 0.8, 'y':
        0.6}, size_hint=(0.2, 0.2), duration=1.5)
        char_animation.start(character_image)

app = TestApp()
app.run()
```

After running the application and pressing the button, the result after the animation ends is shown in Figure 5-6. Note that the duration changes to 1.5 seconds.

Figure 5-6. *Result after animating the pos_hint and size_hint properties of the image*

Clicking the button more does not move or change the size of the image even though the animation still works. In fact, the `start_char_animation()` function is executed after each button press. For each press, an `Animation` instance is created and started according to the attached widget. Talking about the `pos_hint` property for the first time the animation starts, the old value of the `pos_hint` property, specified in the KV file, and the new value, specified inside the `Animation` class constructor, are different. This is why the image moved from x=0.2 to x=0.8. After animating the image, its `pos_hint` property will be `{'x': 0.8, 'y': 0.6}`.

When animating the image again, the start value and the end value will be equal to `{'x': 0.8, 'y': 0.6}`. This is why there is no change in the position of the image widget. Kivy supports looping animations but looping the previous animation makes no sense. There must be at least one other value in the property so that the widget goes from one value to another. Before looping animations, we need to add another value to the `pos_hint` property.

The single animation accepts a single value to a given property, but we can add another value inside another animation and join the animations together.

Joining Animations

There are two ways to join animations—sequential and parallel. In sequential animations, when one animation ends, the next animation starts, and that continues until you reach the last animation. In this case, they are joined using the + operator. In parallel animations, all animations start at the same time. They are joined using the & operator.

Listing 5-10 shows an example in which two animations are joined sequentially. The first animation instance, called char_anim1, moves the image horizontally to the right by changing the pos_hint property to {'x': 0.8, 'y': 0.6}, as was done previously. The second animation instance, called char_anim2, moves the widget vertically to the bottom to the new position, {'x': 0.8, 'y': 0.2}. Both animations are joined using the + operator and the result is stored in the all_anim1 variable. The joined animations start by calling the start() method.

Listing 5-10. Joining Animations Sequentially

```
import kivy.app
import kivy.animation

class TestApp(kivy.app.App):

    def start_char_animation(self):
        character_image = self.root.ids['character_image']
        char_anim1 = kivy.animation.Animation(pos_hint={'x': 0.8, 'y': 0.6})
        char_anim2 = kivy.animation.Animation(pos_hint={'x': 0.8, 'y': 0.2})
        all_anim = char_anim1 + char_anim2
        all_anim.start(character_image)

app = TestApp()
app.run()
```

After pressing the button, the result after running all the animations is shown in Figure 5-7.

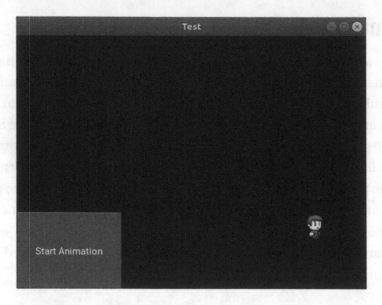

Figure 5-7. *Joining multiple animations to be applied to the Image widget sequentially*

The summary of the path in which the pos_hint property gets changed is illustrated in Figure 5-8. The image starts at {'x': 0.2, 'y': 0.6} specified in the KV file. After running the first animation, it moves to the new position {'x': 0.8, 'y': 0.6}. Finally, it moves to {'x': 0.8, 'y': 0.2} after running the second animation. That position remains the current position of the image.

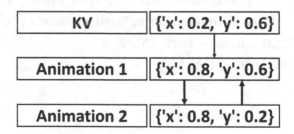

Figure 5-8. *The path of the Image widget according to the two animations defined in Listing 5-10*

After the animation completes, what happens if the button is pressed again? The joined animations will get started again. At the first animation, it changes the position of the image from its current position to the new position specified in its pos_hint argument, which is {'x': 0.8, 'y': 0.6}. Because the current position {'x': 0.8, 'y': 0.2} is different from the new position {'x': 0.8, 'y': 0.6}, the image will move. The current position of the image will be {'x': 0.8, 'y': 0.6}.

After running the first animation, the second animation starts and it moves the image from its current position {'x': 0.8, 'y': 0.6} to the new position specified in its pos_hint argument, which is {'x': 0.8, 'y': 0.2}. Because the positions are different, the image gets moved. This process repeats for each button press. Note that the initial value of the property inside the KV file is lost after the animation, if it's not backed up.

Each animation takes one second to be completed and thus the total time for the joined animation is two seconds. You can control the duration of each animation using the duration argument.

Because there is more than one value by which the pos_hint property changes, we can loop the previous animations. We do this by setting the repeat property of the animation instance to True, according to Listing 5-11. This creates an infinite loop between the two animations.

Listing 5-11. Repeating Animations by Setting the repeat Property to True

```
import kivy.app
import kivy.animation

class TestApp(kivy.app.App):

    def start_char_animation(self):
        character_image = self.root.ids['character_image']
        char_anim1 = kivy.animation.Animation(pos_hint={'x': 0.8, 'y': 0.6})
        char_anim2 = kivy.animation.Animation(pos_hint={'x': 0.8, 'y': 0.2})
        all_anim = char_anim1 + char_anim2
        all_anim.repeat = True
        all_anim.start(character_image)

app = TestApp()
app.run()
```

The animation transition can be changed using the t argument inside the Animation class constructor. The default is linear. There are different types of transitions, such as in_back, in_quad, out_cubic, and many others. You can also return it using the transition property. Listing 5-12 shows an example in which the transition of the first animation is set to out_cubic.

Listing 5-12. Setting the Transition Property to out_cubic

```
import kivy.app
import kivy.animation

class TestApp(kivy.app.App):

    def start_char_animation(self):
        character_image = self.root.ids['character_image']
        char_anim1 = kivy.animation.Animation(pos_hint={'x': 0.8, 'y': 0.6},
        t='out_cubic')
        char_anim2 = kivy.animation.Animation(pos_hint={'x': 0.8, 'y': 0.2})
        all_anim = char_anim1 + char_anim2
        all_anim.repeat = True
        all_anim.start(character_image)

app = TestApp()
app.run()
```

Canceling Animations

In order to stop all animations assigned to all properties in a given widget, we call the cancel_all() function. It stops the all animations once called.

We can add another button to the widget tree that stops all animations when we click it. The new KV file is shown in Listing 5-13. When such a button is pressed, the stop_animation() function is executed. Note that the position of this button has changed, to avoid placing it over the previous button.

Listing 5-13. Adding a Button Widget to Stop Running Animations

```
FloatLayout:
    Image:
        id: character_image
        size_hint: (0.15, 0.15)
        pos_hint: {'x': 0.2, 'y': 0.6}
```

```
        allow_stretch: True
        source: "character.png"
    Button:
        size_hint: (0.3, 0.3)
        text: "Start Animation"
        on_press: app.start_char_animation()
    Button:
        size_hint: (0.3, 0.3)
        text: "Stop Animation"
        pos_hint: {'x': 0.3}
        on_press: app.stop_animation()
```

The Python file is shown in Listing 5-14. Inside the stop_animation() function, the cancel_all() function is called to stop all animations associated with the widget with the character_image ID. When the animations are cancelled, the current values of the animated properties are saved. These values are used as the start values when the animations start again.

Listing 5-14. Stopping Running Animations Upon Press of the Button Widget

```
import kivy.app
import kivy.animation

class TestApp(kivy.app.App):

    def start_char_animation(self):
        character_image = self.root.ids['character_image']
        char_anim1 = kivy.animation.Animation(pos_hint={'x': 0.8, 'y': 0.6})
        char_anim2 = kivy.animation.Animation(pos_hint={'x': 0.8, 'y': 0.2})
        all_anim = char_anim1 + char_anim2
        all_anim.repeat = True
        all_anim.start(character_image)

    def stop_animation(self):
        character_image = self.root.ids['character_image']
        kivy.animation.Animation.cancel_all(character_image)

app = TestApp()
app.run()
```

This way, we are able to start and stop the animations to all properties associated with a given widget. We can also specify selected properties to stop its animation while keeping the others using the `cancel_all()`. Rather than just feeding the widget reference, we add a list of the desired properties to get stopped, separated by commas.

Animating Source Property of the Image Widget

In the previous applications, the same static image is displayed when its position changes. If we want to make the character walk, it is better to change the image as its position changes to give an impression of a walking character. We do this by changing the position of its legs and hands for example. Figure 5-9 shows some images of the character at different positions. While the character is moving, we can also change the image that's displayed. This makes the game more realistic. Because the image is specified using the `source` property inside the `Image` widget, we need to animate this property in order to change the image displayed. The question is how do we animate the `source` property?

Figure 5-9. *Different images to reflect the movement of the character*

In the previous examples, the `pos_hint` and `size_hint` properties are animated and they accept numeric values. But the `source` property accepts a string that specifies the image name. Is it possible to animate a string property? Unfortunately, animations change numeric values only. We can ask the `Animation` class to change a property from a numeric value such as **1.3** to another numeric value such as **5.8**. But we cannot ask it to change a property from a string value such as `character1.png` to another string value, such as `character2.png`. So, how do we do this animation?

One lazy solution consists of four steps. We add a new property to the `Image` widget, say it is named `im_num`, and it will be assigned numbers referring to the index of the image. Then we animate this property to generate the current image number. The third step is to return each value generated by the animation. The last step is to use the generated number to create the image name, by creating a string consisting of the number prepended to the image extension, and set the source property of the Image widget to the returned image name. A summary of this process is illustrated in Figure 5-10. Let's apply these steps.

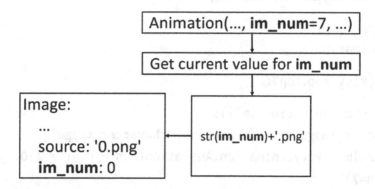

Figure 5-10. *The steps to animate the source property of the Image widget*

For the first step, Listing 5-15 shows the KV file after adding the im_num property. Note that Python allows us to add new properties to already existing classes. The new property is given a value of 0 to refer to the first image of the character.

Listing 5-15. Adding the im_num Property to Change the Image Using Animation

```
FloatLayout:
    Image:
        id: character_image
        size_hint: (0.15, 0.15)
        pos_hint: {'x': 0.2, 'y': 0.6}
        source: "0.png"
        im_num: 0
        allow_stretch: True
    Button:
        size_hint: (0.3, 0.3)
        text: "Start Animation"
        on_press: app.start_char_animation()
```

The second step is simple. We just add an argument named im_num to the constructor of the Animation class. This argument is assigned to the last index to be used. If there are eight images with indices from 0 to 7, this argument is assigned 7. Listing 5-16 shows the Python code.

Listing 5-16. Adding the im_num Argument to the Animation

```python
import kivy.app
import kivy.animation

class TestApp(kivy.app.App):

    def start_char_animation(self):
        character_image = self.root.ids['character_image']
        char_anim = kivy.animation.Animation(pos_hint={'x': 0.8, 'y': 0.6},
        im_num=7)
        char_anim.start(character_image)

app = TestApp()
app.run()
```

In the third step, we have to answer the question, "how do we return the current value generated by the animation?" The answer is simple. To get notified when there is a change to the value of a property named **X** of a given widget, we add an event in that widget named on_X. This event is assigned to a Python function that will be called each time the value of the property changes. Because our target field is named im_num, the event will be called on_im_num.

Listing 5-17 shows the modified KV file after adding this event. Each time the value of the im_num field changes, the function change_char_im() inside the Python file will be called.

Listing 5-17. Adding the on_im_num Event to the Image Widget to Be Notified When the Image Changes

```
FloatLayout:
    Image:
        id: character_image
        size_hint: (0.15, 0.15)
        pos_hint: {'x': 0.2, 'y': 0.6}
        source: "0.png"
        im_num: 0
```

```
        allow_stretch: True
        on_im_num: app.change_char_im()
    Button:
        size_hint: (0.3, 0.3)
        text: "Start Animation"
        on_press: app.start_char_animation()
```

Listing 5-18 shows the modified Python code after adding this function. It is made to print the value of im_num each time it is changed.

Listing 5-18. Handling the on_im_num Event to Print the im_num When Changed

```python
import kivy.app
import kivy.animation

class TestApp(kivy.app.App):

    def change_char_im(self):
        character_image = self.root.ids['character_image']
        print(character_image.im_num)

    def start_char_animation(self):
        character_image = self.root.ids['character_image']
        char_anim = kivy.animation.Animation(pos_hint={'x': 0.8, 'y': 0.6},
        im_num=7)
        char_anim.start(character_image)
app = TestApp()
app.run()
```

In the fourth step, the returned number is concatenated to the image extension to return a string representing the image name. This string is assigned to the source property of the Image module. This work is done inside the modified change_char_im() function, according to Listing 5-19.

Listing 5-19. Changing the Source Image of the Image Widget when im_num
Changes

```
import kivy.app
import kivy.animation

class TestApp(kivy.app.App):

    def change_char_im(self):
        character_image = self.root.ids['character_image']
        character_image.source = str(int(character_image.im_num)) + ".png"

    def start_char_animation(self):
        character_image = self.root.ids['character_image']
        char_anim = kivy.animation.Animation(pos_hint={'x': 0.8, 'y': 0.6},
        im_num=7)
        char_anim.start(character_image)

app = TestApp()
app.run()
```

Note that the animation interpolates floating-point numbers between the starting
and ending values of the animated property. So, there will be values such as 0.1, 2.6, 4.3,
and so on. Because the images have integers in their names, the floating-point value in
the im_num property should be changed to integer.

After converting it into an integer, it can be concatenated with the image extension
to return the string representing the image name. This string is assigned to the source
property of the Image module. Remember to set the images at the current directory of
the Python file. Otherwise, prepend the path to the image name.

After running the application using the latest Python and KV file and pressing the
button, the image should change over time. Figure 5-11 shows four screenshots of the
character while changing its image using animation.

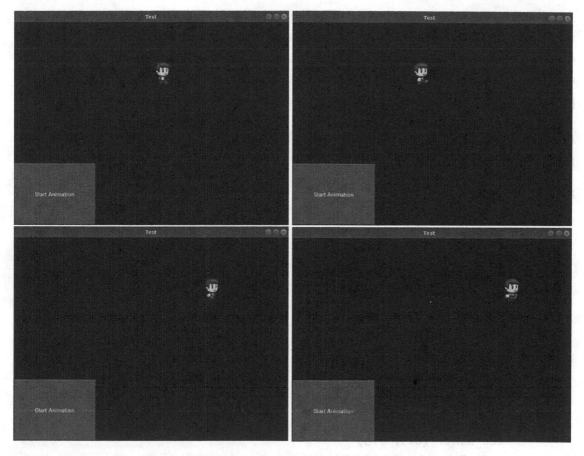

Figure 5-11. *The character image changes while the character is moving*

Screen Touch Events

Up to this point, the character moves when the button is clicked. The movement path is restricted to the one fed into the Animation class constructor. We need to change that in order to move the character freely according to the touch position on the entire screen. Note that touch in Kivy refers to either mouse press or touching a screen. In order to do that, we need to get the touch position on the screen and then animate the character to move to that position.

In order to return the touch position on the screen, there are three touch events to be used, which are on_touch_up, on_touch_down, and on_touch_move. We are just interested in getting the touch position when the touch is down, so the on_touch_down event is used.

This event is added to the root widget (i.e., the FloatLayout) according to the modified KV file in Listing 5-20. Note that binding the touch events with the layout itself or with one of its children does not matter, as they do not detect collisions and thus are unable to detect the boundary of the touch position. They always return the touch position on the entire window.

Listing 5-20. Using the on_touch_down Event to Return the Screen Touch Position

```
FloatLayout:
    on_touch_down: app.touch_down_handler(*args)

    Image:
        id: character_image
        size_hint: (0.15, 0.15)
        pos_hint: {'x': 0.2, 'y': 0.6}
        source: "0.png"
        im_num: 0
        allow_stretch: True
        on_im_num: app.change_char_im()
    Button:
        size_hint: (0.3, 0.3)
        text: "Start Animation"
        on_press: app.start_char_animation()
```

This event accepts a function that will be executed each time the screen is touched. The touch_down_handler() function inside the Python file will be executed in response to the touch. All arguments generated by the event can be passed to the handler using the args variable. This helps to access the touch position inside the Python function.

Listing 5-21 shows the modified Python file that implements the touch_down_handler() function. The function just prints the arguments received from the event in args.

Listing 5-21. Handling the touch_down_handler() to Get the Screen Touch Position

```
import kivy.app
import kivy.animation

class TestApp(kivy.app.App):
```

```python
    def touch_down_handler(self, *args):
        print(args)

    def change_char_im(self):
        character_image = self.root.ids['character_image']
        character_image.source = str(int(character_image.im_num)) + ".png"

    def start_char_animation(self):
        character_image = self.root.ids['character_image']
        char_anim = kivy.animation.Animation(pos_hint={'x': 0.8, 'y': 0.6},
        im_num=7)
        char_anim.start(character_image)

app = TestApp()
app.run()
```

The `args` passed to the function is a tuple with two elements according to the output given. The first element indicates which widget is associated with the event. The second element gives information about the touch and the device fired it. For example, the touch event is fired by a single left mouse button click.

```
(<kivy.uix.floatlayout.FloatLayout object at 0x7fcb70cf4250>,
<MouseMotionEvent button="left" device="mouse" double_tap_time="0"
dpos="(0.0, 0.0)" dsx="0.0" dsy="0.0" dsz="0.0" dx="0.0" dy="0.0"
dz="0.0" grab_current="None" grab_exclusive_class="None" grab_list="[]"
grab_state="False" id="mouse1" is_double_tap="False" is_mouse_
scrolling="False" is_touch="True" is_triple_tap="False" opos="(335.0,
206.99999999999997)" osx="0.45702592087312416" osy="0.37981651376146786"
osz="0.0" ox="335.0" oy="206.99999999999997" oz="0.0" pos="(335.0,
206.99999999999997)" ppos="(335.0, 206.99999999999997)" profile="['pos',
'button']" psx="0.45702592087312416" psy="0.37981651376146786" psz="0.0"
push_attrs="('x', 'y', 'z', 'dx', 'dy', 'dz', 'ox', 'oy', 'oz', 'px', 'py',
'pz', 'pos')" push_attrs_stack="[]" px="335.0" py="206.99999999999997"
pz="0.0" shape="None" spos="(0.45702592087312416, 0.37981651376146786)"
sx="0.45702592087312416" sy="0.37981651376146786" sz="0.0" time_end="-1"
time_start="1563021796.776788" time_update="1563021796.776788" triple_tap_
time="0" ud="{}" uid="1" x="335.0" y="206.99999999999997" z="0.0">)
```

There are different ways by which the event specifies the touch position. For example, the pos property specifies the position in pixels according to the window, while spos returns the position relative to the window size. Because all positions in our game are relative to the window size, the spos is used to specify the position to which the character moves.

Previously, the animation that moves the character was created and started in the start_char_animation() function inside the Python file. This function uses a static position to which the character moves. After using the touch event, the character will move to the position returned in the spos property of the touch event inside the touch_down_handler() function. For that reason, the header of the start_char_animation() function will change to receive the touch position. Listing 5-22 shows the modified Python file.

Note how the spos property is retuned from args. Because it resides in the second element (i.e., index 1) of the args, args[1] is used.

Listing 5-22. Moving the Character According to the Touch Position

```
import kivy.app
import kivy.animation

class TestApp(kivy.app.App):

    def touch_down_handler(self, args):
        self.start_char_animation(args[1].spos)

    def change_char_im(self):
        character_image = self.root.ids['character_image']
        character_image.source = str(int(character_image.im_num)) + ".png"

    def start_char_animation(self, touch_pos):
        character_image = self.root.ids['character_image']
        char_anim = kivy.animation.Animation(pos_hint={'x': touch_pos[0],
        'y': touch_pos[1]}, im_num=7)
        char_anim.start(character_image)

app = TestApp()
app.run()
```

Because the animation now starts by touching the screen, there is no need for the button inside the KV file. Listing 5-23 gives the modified KV file after removing that button.

Listing 5-23. Removing the Button That Starts the Animation

```
FloatLayout:
    on_touch_down: app.touch_down_handler(args)

    Image:
        id: character_image
        size_hint: (0.15, 0.15)
        pos_hint: {'x': 0.2, 'y': 0.6}
        source: "0.png"
        im_num: 0
        allow_stretch: True
        on_im_num: app.change_char_im()
```

After running the application and touching the window, the character will move to the touched position. Because the position of the widget reflects the position at which the bottom-left corner will be placed, feeding that position directly to the pos_hint property of the Image widget makes its bottom-left corner start from the touch position and extends according to its size specified in the size_hint property. This is illustrated in Figure 5-12. It is more convenient to center the widget at the touch position. How do we do that?

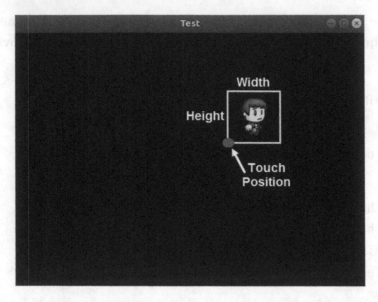

Figure 5-12. *The bottom-left corner of the Image widget is placed on the touch position*

Currently, the center of the widget is larger than the touch position, horizontally by half of its width and vertically by half of its height. The center coordinates are calculated according to the following equations:

```
widgetCenterX = touchPosX + widgetWidth/2
widgetCenterY = touchPosY + widgetHeight/2
```

In order to center the widget according to the touch position, we can subtract widgetWidth/2 from touchPosX and widgetHeight/2 from touchPosY. The result will be as follows:

```
widgetCenterX = (touchPosX - widgetWidth/2) + widgetWidth/2 = touchPosX
widgetCenterY = (touchPosY + widgetHeight/2) + widgetWidth/2 = touchPosY
```

This way, the widget will be centered at the touch position. Listing 5-24 shows the Python code after modifying the position fed to the animation inside the start_ char_animation() function. Note that widgetWidth is equal to size_hint[0] and widgetHeight is equal to size_hint[1].

Listing 5-24. Moving the Center of the Image Widget to the Touch Position

```
import kivy.app
import kivy.animation

class TestApp(kivy.app.App):
    def touch_down_handler(self, args):
        self.start_char_animation(args[1].spos)

    def change_char_im(self):
        character_image = self.root.ids['character_image']
        character_image.source = str(int(character_image.im_num)) + ".png"

    def start_char_animation(self, touch_pos):
        character_image = self.root.ids['character_image']
        char_anim = kivy.animation.Animation(pos_hint={'x': touch_pos[0]-
        character_image.size_hint[0]/2,'y': touch_pos[1]-character_image.
        size_hint[1]/2},im_num=7)
        char_anim.start(character_image)

app = TestApp()
app.run()
```

Figure 5-13 shows the result after touching the screen. The character gets its center positioned at the touch position.

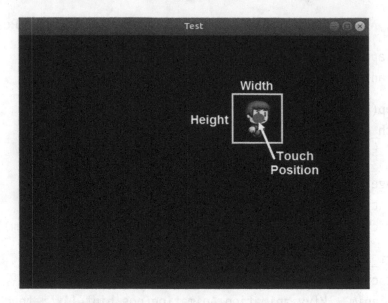

Figure 5-13.   The center of the Image widget is placed on the touch position

on_complete

Each time the screen is touched, an animation instance is created. It animates both the pos_hint and the im_num properties of the Image widget. For the first screen touch, both properties are animated as the character moves and the image gets changed. Touching the screen more only moves the character but, unfortunately, the im_num property does not change. Thus, there will be a static image displayed on the widget after the first screen touch. Why does this happen?

The initial value of the im_num inside the KV file is 0. When the screen is touched for the first time, the animation starts and thus the im_num animates from 0 to 7. After the animation completes, the current value stored in the im_num will be 7.

Touching the screen another time, the im_num will animate from the current value 7 to the new value, which is also 7. As a result, there is no change to the displayed image. The solution is to reset the value of im_num to 0 before starting the animation inside the start_char_animation() function.

After the animation is completed, the character is expected to be in a stable state and thus the first image with im_num=0 is to be displayed. But the value stored in im_num after completing the animation is 7, not 0. It would be better to reset im_num to 0 after the animation is complete.

Fortunately, the Animation class has an event called on_complete that is fired when the animation completes. We can bind our animation to this event so that a callback function is executed each time it is completed. The callback function is named char_animation_completed(). Inside that function, we can force im_num to return to 0. The modified Python file is listed in Listing 5-25 after binding the on_complete event to the animation of the character. This on_complete event sends arguments to the callback function, which are the animation firing the event and the widget associated with it. This is why the callback function accepts them into args.

Listing 5-25. Resetting the im_num Property to 0 After the Animation Completes

```python
import kivy.app
import kivy.animation

class TestApp(kivy.app.App):

    def char_animation_completed(self, *args):
        character_image = self.root.ids['character_image']
        character_image.im_num = 0

    def touch_down_handler(self, args):
        self.start_char_animation(args[1].spos)

    def change_char_im(self):
        character_image = self.root.ids['character_image']
        character_image.source = str(int(character_image.im_num))+".png"

    def start_char_animation(self, touch_pos):
        character_image = self.root.ids['character_image']
        character_image.im_num = 0
        char_anim = kivy.animation.Animation(pos_hint={'x': touch_pos[0]-
        character_image.size_hint[0]/2, 'y': touch_pos[1]-character_image.
        size_hint[1]/2}, im_num=7)
        char_anim.bind(on_complete=self.char_animation_completed)
        char_anim.start(character_image)

app = TestApp()
app.run()
```

153

Inside the callback function, the im_num of the Image widget is changed back to 0. As a result, the image displayed on the widget will be reset each time the animation completes.

After animating the character correctly, we can start adding a monster to the game. The player/character dies when it collides with the monster.

Adding Monsters to the Game

The monster will be added to the KV file the same way the character is added. We just create an Image widget in the KV file for the monster. The new KV file is shown in Listing 5-26.

Listing 5-26. Adding an Image Widget for the Monster

```
FloatLayout:
    on_touch_down: app.touch_down_handler(args)

    Image:
        id: monster_image
        size_hint: (0.15, 0.15)
        pos_hint: {'x': 0.8, 'y': 0.8}
        source: "10.png"
        im_num: 10
        allow_stretch: True
        on_im_num: app.change_monst_im()

    Image:
        id: character_image
        size_hint: (0.15, 0.15)
        pos_hint: {'x': 0.2, 'y': 0.6}
        source: "0.png"
        im_num: 0
        allow_stretch: True
        on_im_num: app.change_char_im()
```

The Image widget of the monster will have the properties defined in the character. The properties are an ID for referring to the widget inside the Python file, size_hint for setting the widget size relative to the screen size, pos_hint to place the widget relative to the screen, source for holding the image name as a string, allow_stretch to stretch the image to cover the entire area of the Image, and im_num to hold the image number displayed on the widget. To make the image numbers of the character and the monster different, the monster image numbers will start with 10. Figure 5-14 shows the images of the monster.

Figure 5-14. *The image sequence of the monster*

The on_im_num event will be associated with the im_num property with a callback function named change_monst_im() to gain access to its value inside the Python file each time it changed.

After we prepare the KV file, the application window we would see is shown in Figure 5-15.

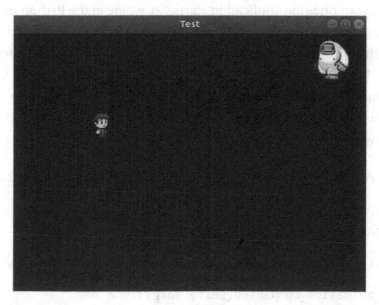

Figure 5-15. *The character and the monster Image widgets appear on the screen*

155

Note that the monster Image widget is positioned before the character widget in the KV file (i.e., in the widget tree). This makes the character Z index lower than the Z index of the monster and thus drawn above it, as shown in Figure 5-16.

Figure 5-16. *The character Image widget appears above of the monster Image widget*

on_start

The animation of the character starts each time there is a screen touch but the monster animation must start once the application starts. So, where in the Python file we can start the monster? According to the Kivy application lifecycle, the method called on_start() is executed once the application starts. This is a good place to start the monster's animation.

The Python file, after we add the change_monst_im() and on_start() functions to handle the animation of the monster, is shown in Listing 5-27. The change_monst_im() function is similar to change_char_im(), except for changing the source property of the monster Image widget.

Listing 5-27. Adding the Functions to Handle the Monster Animation

```
import kivy.app
import kivy.animation

class TestApp(kivy.app.App):

    def char_animation_completed(self, *args):
        character_image = self.root.ids['character_image']
        character_image.im_num = 0
```

```python
    def touch_down_handler(self, args):
        self.start_char_animation(args[1].spos)

    def change_char_im(self):
        character_image = self.root.ids['character_image']
        character_image.source = str(int(character_image.im_num))+".png"

    def change_monst_im(self):
        monster_image = self.root.ids['monster_image']
        monster_image.source = str(int(monster_image.im_num))+".png"

    def start_char_animation(self, touch_pos):
        character_image = self.root.ids['character_image']
        character_image.im_num = 0
        char_anim = kivy.animation.Animation(pos_hint={'x': touch_pos[0]-
        character_image.size_hint[0]/2, 'y': touch_pos[1]-character_image.
        size_hint[1]/2}, im_num=7)
        char_anim.bind(on_complete=self.char_animation_completed)
        char_anim.start(character_image)

    def on_start(self):
        monster_image = self.root.ids['monster_image']
        monst_anim = kivy.animation.Animation(pos_hint={'x': 0.8, 'y': 0.0},
        im_num=17, duration=2.0)+kivy.animation.Animation(pos_hint={'x': 0.8,
        'y': 0.8}, im_num=10, duration=2.0)
        monst_anim.repeat = True
        monst_anim.start(monster_image)

app = TestApp()
app.run()
```

Inside the on_start() function, two sequentially joined animations are created to animate the pos_hint and im_num properties of the monster Image widget.

The initial position of the monster according to the KV file is {'x':0.8, 'y':0.8}. The first animation changes that position to {'x':0.8, 'y':0.0} and the second animation changes it back to {'x':0.8, 'y':0.8}. This animation takes place in a loop because the repeat property of the animation instance monst_anim is set to True. The monster moves in a fixed path for simplicity. In the next sections, we will change its movement to be random.

Because the initial value of the im_num property of the monster in the KV file is set to 10, the first animation has im_num set to 17. As a result, the first animation changes the image number from 10 to 17. The second animation has such a property set to 10 to change the image number from 17 back to 10. Each animation lasts two seconds.

The animation is associated with the monster Image widget, which is returned using its ID inside the monster_image variable.

Collision

Up to this point, the animations for both the pos_hint and im_num properties of the character and monster are working fine. We need to modify the game so that the character is killed when it collides with the monster.

The way collision works in Kivy is that it checks for an intersection between the bounding boxes of two widgets. The collide_widget() built-in Kivy function does that. For example, this command detects the collision between the two Image widgets:

```
character_image.collide_widget(monster_image)
```

We have to continuously check for collision between the two widgets. Thus, the above command needs to be added to something that is periodically executed.

Each time a widget changes its position using the pos_hint property, the on_pos_hint event is fired. This event will execute a callback function each time it is fired. Because the monster Image widget continuously changes its position, we can bind the event to that widget.

Note that if you are intending to kill the monster later, the monster will not change its position and thus the on_pos_hint will never be fired and thus there will be no check for collision. If there are other objects that may kill the character and you completely depend on the event associated with the killed monster for collision detection, the character will not be killed. You have to find something else to check the collision. One solution is to bind the on_pos_hint event with each object that might kill the character.

At first, the on_pos_hint event is added to the KV file, as given in Listing 5-28. It is associated with a callback function called monst_pos_hint().

Listing 5-28. Using the on_pos_hint Event to Return the Monster Position

```
FloatLayout:
    on_touch_down: app.touch_down_handler(args)

    Image:
        id: monster_image
        size_hint: (0.15, 0.15)
        pos_hint: {'x': 0.8, 'y': 0.8}
        source: "10.png"
        im_num: 10
        allow_stretch: True
        on_im_num: app.change_monst_im()
        on_pos_hint: app.monst_pos_hint()

    Image:
        id: character_image
        size_hint: (0.15, 0.15)
        pos_hint: {'x': 0.2, 'y': 0.6}
        source: "0.png"
        im_num: 0
        allow_stretch: True
        on_im_num: app.change_char_im()
```

The monst_pos_hint() function is implemented at the end of the Python file shown in Listing 5-29. It returns the character and monster widgets in both the character_image and monster_image properties and then calls the collide_widget() function. Just up to this time, a message is printed if a collision has occurred according to an if statement.

Listing 5-29. Handling the monst_pos_hint() Callback Function

```
import kivy.app
import kivy.animation

class TestApp(kivy.app.App):

    def char_animation_completed(self, *args):
        character_image = self.root.ids['character_image']
        character_image.im_num = 0
```

```python
    def touch_down_handler(self, args):
        self.start_char_animation(args[1].spos)

    def change_char_im(self):
        character_image = self.root.ids['character_image']
        character_image.source = str(int(character_image.im_num))+".png"

    def change_monst_im(self):
        monster_image = self.root.ids['monster_image']
        monster_image.source = str(int(monster_image.im_num))+".png"

    def start_char_animation(self, touch_pos):
        character_image = self.root.ids['character_image']
        character_image.im_num = 0
        char_anim = kivy.animation.Animation(pos_hint={'x': touch_pos[0]-
        character_image.size_hint[0]/2, 'y': touch_pos[1]-character_image.
        size_hint[1]/2}, im_num=7)
        char_anim.bind(on_complete=self.char_animation_completed)
        char_anim.start(character_image)

    def on_start(self):
        monster_image = self.root.ids['monster_image']
        monst_anim = kivy.animation.Animation(pos_hint={'x': 0.8, 'y':
        0.0}, im_num=17, duration=2.0)+kivy.animation.Animation(pos_
        hint={'x': 0.8, 'y': 0.8}, im_num=10, duration=2.0)
        monst_anim.repeat = True
        monst_anim.start(monster_image)

    def monst_pos_hint(self):
        character_image = self.root.ids['character_image']
        monster_image = self.root.ids['monster_image']
        if character_image.collide_widget(monster_image):
            print("Character Killed")

app = TestApp()
app.run()
```

Tuning collide_widget()

The collide_widget() function is too strict, as it returns True if there is at least an intersection in a single row or column between the two widgets. In practice, this does not happen frequently. According to Figure 5-17, such a function returns True because there is intersection between the bounding boxes of the two widgets. As a result, the character will be killed even if it hasn't touched the monster.

Figure 5-17. *The collide_widget() returns True when the outer boundary of the widget boxes collide, even if the images have not touched*

We can tune the collide_widget() function by adding another condition that checks whether the collision area exceeds a predefined percentage of the character size. This makes us more confident when saying that there is collision. The monst_pos_hint() function is modified as shown in Listing 5-30.

Listing 5-30. Tuning the collide_widget() Function to Return True Only When the Character and Monster Touch Each Other

```
def monst_pos_hint(self):
    character_image = self.root.ids['character_image']
    monster_image = self.root.ids['monster_image']

    character_center = character_image.center
    monster_center = monster_image.center

    gab_x = character_image.width / 2
    gab_y = character_image.height / 2
```

```
if character_image.collide_widget(monster_image) and abs(character_
center[0] - monster_center[0]) <= gab_x and abs(character_center[1] -
monster_center[1]) <= gab_y:
    print("Character Killed")
```

The new condition concludes that collision occurs if the difference between the current centers of the two widgets is at least one half of the character size. This is done by making sure that the difference between the X and Y coordinates of the two centers is less than one half of the character width and height, respectively.

After tuning the `collide_widget()` function, the condition returned `False` for the cases presented in Figures 5-18 and 5-19. Thus, the results are more realistic.

Figure 5-18. *No collision occurred because the maximum gab defined in Listing 5-30 between the center of the monster and character images is not exceeded. The complete Python code is shown in Listing 5-31*

Listing 5-31. Complete Code for the Game in Which the Work for the Animation and Collision Is Completed Successfully

```
import kivy.app
import kivy.animation

class TestApp(kivy.app.App):

    def char_animation_completed(self, *args):
        character_image = self.root.ids['character_image']
        character_image.im_num = 0

    def touch_down_handler(self, args):
        self.start_char_animation(args[1].spos)

    def change_char_im(self):
        character_image = self.root.ids['character_image']
        character_image.source = str(int(character_image.im_num))+".png"
```

```python
    def change_monst_im(self):
        monster_image = self.root.ids['monster_image']
        monster_image.source = str(int(monster_image.im_num))+".png"

    def start_char_animation(self, touch_pos):
        character_image = self.root.ids['character_image']
        character_image.im_num = 0
        char_anim = kivy.animation.Animation(pos_hint={'x': touch_pos[0]-
        character_image.size_hint[0]/2, 'y': touch_pos[1]-character_image.
        size_hint[1]/2}, im_num=7)
        char_anim.bind(on_complete=self.char_animation_completed)
        char_anim.start(character_image)

    def on_start(self):
        monster_image = self.root.ids['monster_image']
        monst_anim = kivy.animation.Animation(pos_hint={'x': 0.8, 'y':
        0.0}, im_num=17, duration=2.0)+kivy.animation.Animation(pos_
        hint={'x': 0.8, 'y': 0.8}, im_num=10, duration=2.0)
        monst_anim.repeat = True
        monst_anim.start(monster_image)

    def monst_pos_hint(self):
        character_image = self.root.ids['character_image']
        monster_image = self.root.ids['monster_image']

        character_center = character_image.center
        monster_center = monster_image.center

        gab_x = character_image.width/2
        gab_y = character_image.height/2
        if character_image.collide_widget(monster_image) and abs(character_
        center[0]-monster_center[0])<=gab_x and abs(character_center[1]-
        monster_center[1])<=gab_y:
            print("Character Killed")

app = TestApp()
app.run()
```

Random Monster Motion

In the previous application in Listing 5-31, we did a good job of animating the character and the monster successfully. But the monster moves in a fixed path. In this section, we will modify its motion to make it seem random. The idea is very similar to how the character Image widget is animated.

The character is animated using a function called `start_char_animation()`, which accepts the new position to which the character moves. Because there is just a single animation created, it can be repeated. In order to repeat the animation after being completed, the `on_complete` event is attached to the animation of the character. A callback function named `char_animation_completed()` is associated with the event. When the animation completes, this callback function is executed where it prepares the character for the new animation. We would like to make the movement of the monster similar to that. This is applied to the modified Python file shown in Listing 5-32.

Two new functions are created, which are `start_char_animation()` and `char_animation_completed()`. The `start_char_animation()` function accepts the new position to which the monster moves as an argument named `new_pos`. Then it creates an animation instance that changes the `pos_hint` property according to the new position. It also changes the `im_num` property from the initial value of 10 in the KV file to 17.

Listing 5-32. Repeating the Monster Animation by Handling the on_complete Event of the Animation

```python
import kivy.app
import kivy.animation
import random

class TestApp(kivy.app.App):

    def char_animation_completed(self, *args):
        character_image = self.root.ids['character_image']
        character_image.im_num = 0

    def monst_animation_completed(self, *args):
        monster_image = self.root.ids['monster_image']
        monster_image.im_num = 10
```

```python
    new_pos = (random.uniform(), random.uniform())
    self.start_monst_animation(new_pos=new_pos)

def touch_down_handler(self, args):
    self.start_char_animation(args[1].spos)

def change_char_im(self):
    character_image = self.root.ids['character_image']
    character_image.source = str(int(character_image.im_num)) + ".png"

def change_monst_im(self):
    monster_image = self.root.ids['monster_image']
    monster_image.source = str(int(monster_image.im_num)) + ".png"

def start_char_animation(self, touch_pos):
    character_image = self.root.ids['character_image']
    character_image.im_num = 0
    char_anim = kivy.animation.Animation(pos_hint={'x': touch_pos[0] -
    character_image.size_hint[0] / 2, 'y': touch_pos[1] - character_
    image.size_hint[1] / 2}, im_num=7)
    char_anim.bind(on_complete=self.char_animation_completed)
    char_anim.start(character_image)

def start_monst_animation(self, new_pos, anim_duration):
    monster_image = self.root.ids['monster_image']
    monst_anim = kivy.animation.Animation(pos_hint={'x': new_pos[0],
    'y': new_pos[1]}, im_num=17, duration=anim_duration)
    monst_anim.bind(on_complete=self.monst_animation_completed)
    monst_anim.start(monster_image)

def on_start(self):
    monster_image = self.root.ids['monster_image']
    new_pos = (random.uniform(0.0, 1 - monster_image.size_hint[0]),
    random.uniform(0.0, 1 - monster_image.size_hint[1]))
    self.start_monst_animation(new_pos=new_pos, anim_duration=random.
    uniform(1.5, 3.5))
```

```
    def monst_pos_hint(self):
        character_image = self.root.ids['character_image']
        monster_image = self.root.ids['monster_image']

        character_center = character_image.center
        monster_center = monster_image.center

        gab_x = character_image.width / 2
        gab_y = character_image.height / 2
        if character_image.collide_widget(monster_image) and abs(character_
        center[0] - monster_center[0]) <= gab_x and abs(character_
        center[1] - monster_center[1]) <= gab_y:
            print("Character Killed")

app = TestApp()
app.run()
```

Inside the on_start() function, the start_monst_animation() is called with the input argument specified to a random value. Because there is just a single animation, the animation cannot repeat itself by setting the repeat property to True. For that reason, the on_complete event is attached to the animation so that the callback function monst_animation_completed() is executed after the animation is completed. This gives us the chance to start the animation again.

Inside the callback function, the im_num property of the monster is reset to 10 again. Using the uniform() function inside the random module, a random value is generated for the X and Y coordinates of the new position. The returned value is a floating-point number between 0.0 and 1.0. The new position is used as the bottom-left corner of the monster widget.

Assuming that the randomly returned position is (0.0, 1.0), this makes the character's bottom line start at the end of the screen. As a result, the monster will be hidden. This also occurs for the positions (1.0, 0.0) and (1.0, 1.0).

In order to make sure that the monster is always visible in the screen, we have to take its width and height in regard. After positioning the bottom-left corner of the monster at the new random position, the monster there must be a space for its width and height. Thus, the maximum possible value for X is 1-monster_width and the maximum possible Y value is 1-monster_height. This makes room for the monster to be completely visible in the window for any generated position.

The modified Python code is shown in Listing 5-33. In the previous application, the duration for all monster movements is 2.0 seconds. In the new code, the duration is randomly returned using `random.uniform()`. As a result, the monster moves to randomly generated positions in random durations.

Listing 5-33. Randomly Changing the Position and Duration of the Monster Animation

```python
import kivy.app
import kivy.animation
import random

class TestApp(kivy.app.App):

    def char_animation_completed(self, *args):
        character_image = self.root.ids['character_image']
        character_image.im_num = 0

    def monst_animation_completed(self, *args):
        monster_image = self.root.ids['monster_image']
        monster_image.im_num = 10

        new_pos = (random.uniform(0.0, 1 - monster_image.size_hint[0]),
        random.uniform(0.0, 1 - monster_image.size_hint[1]))
        self.start_monst_animation(new_pos= new_pos,anim_duration=random.
        uniform(1.5, 3.5))

    def touch_down_handler(self, args):
        self.start_char_animation(args[1].spos)

    def change_char_im(self):
        character_image = self.root.ids['character_image']
        character_image.source = str(int(character_image.im_num)) + ".png"

    def change_monst_im(self):
        monster_image = self.root.ids['monster_image']
        monster_image.source = str(int(monster_image.im_num)) + ".png"

    def start_char_animation(self, touch_pos):
        character_image = self.root.ids['character_image']
        character_image.im_num = 0
```

```
        char_anim = kivy.animation.Animation(pos_hint={'x': touch_pos[0] -
        character_image.size_hint[0] / 2, 'y': touch_pos[1] - character_
        image.size_hint[1] / 2}, im_num=7)
        char_anim.bind(on_complete=self.char_animation_completed)
        char_anim.start(character_image)

    def start_monst_animation(self, new_pos, anim_duration):
        monster_image = self.root.ids['monster_image']
        monst_anim = kivy.animation.Animation(pos_hint={'x': new_pos[0],
        'y': new_pos[1]}, im_num=17, duration=anim_duration)
        monst_anim.bind(on_complete=self.monst_animation_completed)
        monst_anim.start(monster_image)

    def on_start(self):
        monster_image = self.root.ids['monster_image']
        new_pos = (random.uniform(0.0, 1 - monster_image.size_hint[0]),
        random.uniform(0.0, 1 - monster_image.size_hint[1]))
        self.start_monst_animation(new_pos=new_pos, anim_duration=random.
        uniform(1.5, 3.5))

    def monst_pos_hint(self):
        character_image = self.root.ids['character_image']
        monster_image = self.root.ids['monster_image']

        character_center = character_image.center
        monster_center = monster_image.center

        gab_x = character_image.width / 2
        gab_y = character_image.height / 2
        if character_image.collide_widget(monster_image) and abs(character_
        center[0] - monster_center[0]) <= gab_x and abs(character_
        center[1] - monster_center[1]) <= gab_y:
            print("Character Killed")

app = TestApp()
app.run()
```

Killing the Character

In the previous game in Listing 5-33, everything still works even after the monster collides with the character. We need to modify the application so that the character stops moving when it's killed. We do this by making sure that its animation will not start again.

Inside the touch_down_handler() function, the character is always moving toward the touched position on the screen, even after collision. In the modified Python code listed in Listing 5-34, this is fixed that by using a flag variable named character_killed indicating whether the character is killed or not. Such a variable is set to False by default, meaning that the game is still running and the character is alive. An if statement inside the touch_down_handler() function makes sure that the character animation only works when that flag is set to False. Because the flag is associated with the class, it can be accessed by prepending the class name (TestApp.character_killed).

When a collision is detected inside the mons_pos_hint() function, two actions are taken, which are changing the value of the character_killed flag to True and canceling all running animations (i.e., character and monster).

Listing 5-34. Adding the character_killed Flag to Determine Whether the Character Animation Could Start Again

```python
import kivy.app
import kivy.animation
import random

class TestApp(kivy.app.App):
    character_killed = False

    def char_animation_completed(self, *args):
        character_image = self.root.ids['character_image']
        character_image.im_num = 0

    def monst_animation_completed(self, *args):
        monster_image = self.root.ids['monster_image']
        monster_image.im_num = 10

        new_pos = (random.uniform(0.0, 1 - monster_image.size_hint[0]),
        random.uniform(0.0, 1 - monster_image.size_hint[1]))
        self.start_monst_animation(new_pos= new_pos,anim_duration=random.
        uniform(1.5, 3.5))
```

```
def touch_down_handler(self, args):
    if TestApp.character_killed == False:
        self.start_char_animation(args[1].spos)

def change_char_im(self):
    character_image = self.root.ids['character_image']
    character_image.source = str(int(character_image.im_num)) + ".png"

def change_monst_im(self):
    monster_image = self.root.ids['monster_image']
    monster_image.source = str(int(monster_image.im_num)) + ".png"

def start_char_animation(self, touch_pos):
    character_image = self.root.ids['character_image']
    character_image.im_num = 0
    char_anim = kivy.animation.Animation(pos_hint={'x': touch_pos[0] -
    character_image.size_hint[0] / 2, 'y': touch_pos[1] - character_
    image.size_hint[1] / 2}, im_num=7)
    char_anim.bind(on_complete=self.char_animation_completed)
    char_anim.start(character_image)

def start_monst_animation(self, new_pos, anim_duration):
    monster_image = self.root.ids['monster_image']
    monst_anim = kivy.animation.Animation(pos_hint={'x': new_pos[0],
    'y': new_pos[1]}, im_num=17, duration=anim_duration)
    monst_anim.bind(on_complete=self.monst_animation_completed)
    monst_anim.start(monster_image)

def on_start(self):
    monster_image = self.root.ids['monster_image']
    new_pos = (random.uniform(0.0, 1 - monster_image.size_hint[0]),
    random.uniform(0.0, 1 - monster_image.size_hint[1]))
    self.start_monst_animation(new_pos=new_pos, anim_duration=random.
    uniform(1.5, 3.5))

def monst_pos_hint(self):
    character_image = self.root.ids['character_image']
    monster_image = self.root.ids['monster_image']
```

170

```
character_center = character_image.center
monster_center = monster_image.center

gab_x = character_image.width / 2
gab_y = character_image.height / 2
if character_image.collide_widget(monster_image) and abs(character_
center[0] - monster_center[0]) <= gab_x and abs(character_
center[1] - monster_center[1]) <= gab_y:

    kivy.animation.Animation.cancel_all(character_image)
    kivy.animation.Animation.cancel_all(monster_image)

app = TestApp()
app.run()
```

When the flag changes to True inside the monst_pos_hint() function, the character animation cannot be stopped. Note that after the flag value changes to True there is still a running animation in response to the previously touched position. This means that the character will continue moving until the animation completes and then will stop moving after that. In order to stop moving the animation once the collision occurs, we can cancel the animation using the cancel_all() function. As a result, cancelling the animation will stop it once collision takes place. Changing the flag value stops the animation from getting started again.

Because there is no way for the user to start the monster animation once it has been cancelled, it is enough to cancel such an animation.

Animation on Character Kill

When the character is killed in the previous application, it holds on the image whose number is specified by the im_num property, according to Figure 5-19. The image does not reflect the character's death.

Figure 5-19. *The character image stops at its latest state when collision occurs with the monster*

We can change the image to give a better impression. For that purpose, the images displayed in Figure 5-20 will be used.

Figure 5-20. *Sequence of images to be displayed when the character is killed*

Once collision occurs, it will start only after the animation is created inside the `monst_pos_hint()` function that animates the image according to such images. If these images are numbered from 91 to 95, the modified Python code is shown in Listing 5-35. The new animation just changes the `im_num` property to 95.

It is important to remember that after the character animation is cancelled, the `im_num` number will hold on a number between 0 and 7. If, for example, its value was 5, then running the new animation will go from 5 to 95. Because we are interested in starting at 91, the `im_num` property value is set to 91 before the animation starts.

Listing 5-35. Running an Animation When the Character Is Killed

```
import kivy.app
import kivy.animation
import random

class TestApp(kivy.app.App):
    character_killed = False
```

```python
def char_animation_completed(self, *args):
    character_image = self.root.ids['character_image']
    character_image.im_num = 0

def monst_animation_completed(self, *args):
    monster_image = self.root.ids['monster_image']
    monster_image.im_num = 10

    new_pos = (random.uniform(0.0, 1 - monster_image.size_hint[0]),
    random.uniform(0.0, 1 - monster_image.size_hint[1]))
    self.start_monst_animation(new_pos= new_pos,anim_duration=random.
    uniform(1.5, 3.5))

def touch_down_handler(self, args):
    if TestApp.character_killed == False:
        self.start_char_animation(args[1].spos)

def change_char_im(self):
    character_image = self.root.ids['character_image']
    character_image.source = str(int(character_image.im_num)) + ".png"

def change_monst_im(self):
    monster_image = self.root.ids['monster_image']
    monster_image.source = str(int(monster_image.im_num)) + ".png"

def start_char_animation(self, touch_pos):
    character_image = self.root.ids['character_image']
    character_image.im_num = 0
    char_anim = kivy.animation.Animation(pos_hint={'x': touch_pos[0] -
    character_image.size_hint[0] / 2, 'y': touch_pos[1] - character_
    image.size_hint[1] / 2}, im_num=7)
    char_anim.bind(on_complete=self.char_animation_completed)
    char_anim.start(character_image)

def start_monst_animation(self, new_pos, anim_duration):
    monster_image = self.root.ids['monster_image']
    monst_anim = kivy.animation.Animation(pos_hint={'x': new_pos[0],
    'y': new_pos[1]}, im_num=17, duration=anim_duration)
    monst_anim.bind(on_complete=self.monst_animation_completed)
    monst_anim.start(monster_image)
```

173

```python
    def on_start(self):
        monster_image = self.root.ids['monster_image']
        new_pos = (random.uniform(0.0, 1 - monster_image.size_hint[0]),
        random.uniform(0.0, 1 - monster_image.size_hint[1]))
        self.start_monst_animation(new_pos=new_pos, anim_duration=random.
        uniform(1.5, 3.5))

    def monst_pos_hint(self):
        character_image = self.root.ids['character_image']
        monster_image = self.root.ids['monster_image']

        character_center = character_image.center
        monster_center = monster_image.center

        gab_x = character_image.width / 2
        gab_y = character_image.height / 2
        if character_image.collide_widget(monster_image) and abs(character_
        center[0] - monster_center[0]) <= gab_x and abs(character_
        center[1] - monster_center[1]) <= gab_y:
            TestApp.character_killed = True

            kivy.animation.Animation.cancel_all(character_image)
            kivy.animation.Animation.cancel_all(monster_image)

            character_image.im_num = 91
            char_anim = kivy.animation.Animation(im_num=95)
            char_anim.start(character_image)

app = TestApp()
app.run()
```

According to the code in Listing 5-35, the result will be as shown in Figure 5-21 when collision occurs.

Figure 5-21. *The character image changes when it collides with the monster*

Adding Coins

The mission of the player is to collect a number of coins distributed across the screen. Once the right number of coins is collected, the current level of the game is completed and another level starts. Thus, the next step in the application is to add Image widgets in the widget tree representing coins. Let`s start by adding just a single Image widget representing one coin.

According to the Kivy application lifecycle, the build() method is available to prepare the widget tree. Thus, it is a good way to add new widgets to the application. The Python code shown in Listing 5-36 implements the build() method to add a single Image widget. Remember to import the kivy.uix.image module in order to access the Image class.

Listing 5-36. Adding an Image Widget to the Widget Tree Representing the Coin Before the Application Starts

```
import kivy.app
import kivy.animation
import kivy.uix.image
import random

class TestApp(kivy.app.App):
    character_killed = False

    def char_animation_completed(self, *args):
        character_image = self.root.ids['character_image']
        character_image.im_num = 0
```

```python
def monst_animation_completed(self, *args):
    monster_image = self.root.ids['monster_image']
    monster_image.im_num = 10

    new_pos = (random.uniform(0.0, 1 - monster_image.size_hint[0]),
    random.uniform(0.0, 1 - monster_image.size_hint[1]))
    self.start_monst_animation(new_pos= new_pos,anim_duration=random.
    uniform(1.5, 3.5))

def touch_down_handler(self, args):
    if TestApp.character_killed == False:
        self.start_char_animation(args[1].spos)

def change_char_im(self):
    character_image = self.root.ids['character_image']
    character_image.source = str(int(character_image.im_num)) + ".png"

def change_monst_im(self):
    monster_image = self.root.ids['monster_image']
    monster_image.source = str(int(monster_image.im_num)) + ".png"

def start_char_animation(self, touch_pos):
    character_image = self.root.ids['character_image']
    character_image.im_num = 0
    char_anim = kivy.animation.Animation(pos_hint={'x': touch_pos[0] -
    character_image.size_hint[0] / 2, 'y': touch_pos[1] - character_
    image.size_hint[1] / 2}, im_num=7)
    char_anim.bind(on_complete=self.char_animation_completed)
    char_anim.start(character_image)

def start_monst_animation(self, new_pos, anim_duration):
    monster_image = self.root.ids['monster_image']
    monst_anim = kivy.animation.Animation(pos_hint={'x': new_pos[0],
    'y': new_pos[1]}, im_num=17, duration=anim_duration)
    monst_anim.bind(on_complete=self.monst_animation_completed)
    monst_anim.start(monster_image)
```

```python
def build(self):
    coin = kivy.uix.image.Image(source="coin.png", size_hint=(0.05,
    0.05), pos_hint={'x': 0.5, 'y': 0.5}, allow_stretch=True)
    self.root.add_widget(coin, index=-1)

def on_start(self):
    monster_image = self.root.ids['monster_image']
    new_pos = (random.uniform(0.0, 1 - monster_image.size_hint[0]),
    random.uniform(0.0, 1 - monster_image.size_hint[1]))
    self.start_monst_animation(new_pos=new_pos, anim_duration=random.
    uniform(1.5, 3.5))

def monst_pos_hint(self):
    character_image = self.root.ids['character_image']
    monster_image = self.root.ids['monster_image']

    character_center = character_image.center
    monster_center = monster_image.center

    gab_x = character_image.width / 2
    gab_y = character_image.height / 2
    if character_image.collide_widget(monster_image) and abs(character_
    center[0] - monster_center[0]) <= gab_x and abs(character_
    center[1] - monster_center[1]) <= gab_y:
        TestApp.character_killed = True

        kivy.animation.Animation.cancel_all(character_image)
        kivy.animation.Animation.cancel_all(monster_image)

        character_image.im_num = 91
        char_anim = kivy.animation.Animation(im_num=95)
        char_anim.start(character_image)

        print("Character Killed")

app = TestApp()
app.run()
```

The new widget uses the source, size_hint, pos_hint, and allow_stretch properties. The coin image source is shown in Figure 5-22.

Figure 5-22. *Coin image*

The new widget is returned to the `coin` variable. After that, it is added to the widget tree using the `add_widget()` method. Because the last widget in the widget tree appears on the top of the previous widgets, the `index` argument is used to alter the Z index of the coin. The default Z index for widgets is 0. The coin Z index is set to -1 to appear behind both the character and the monster.

After we run the application, we'll see the window shown in Figure 5-23. We can add more coins to the window.

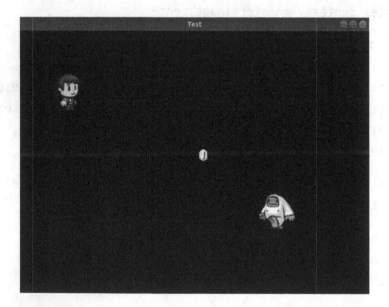

Figure 5-23. *Adding the coin next to the character and monster Image widgets*

One way of adding coins is to fix their positions on the screen. In this game, the positions will be random. The modified `build()` function is shown in Listing 5-37 in which a `for` loop adds five coin Image widgets to the widget tree. Note that there is a variable called num_coins defined in the class header and it holds the number of coin widgets.

The uniform() function is used to return the x and y coordinates of each coin, taking into regard the leaving place for displaying the coin on the screen. We do this is by subtracting the width and height from the returned random number. This is the same way the monster's random position is generated.

Listing 5-37. Adding Multiple Image Widgets Representing the Coins on the Screen

```python
import kivy.app
import kivy.animation
import kivy.uix.image
import random

class TestApp(kivy.app.App):
    character_killed = False
    num_coins = 5

    def char_animation_completed(self, *args):
        character_image = self.root.ids['character_image']
        character_image.im_num = 0

    def monst_animation_completed(self, *args):
        monster_image = self.root.ids['monster_image']
        monster_image.im_num = 10

        new_pos = (random.uniform(0.0, 1 - monster_image.size_hint[0]),
        random.uniform(0.0, 1 - monster_image.size_hint[1]))
        self.start_monst_animation(new_pos= new_pos,anim_duration=random.
        uniform(1.5, 3.5))

    def touch_down_handler(self, args):
        if TestApp.character_killed == False:
            self.start_char_animation(args[1].spos)

    def change_char_im(self):
        character_image = self.root.ids['character_image']
        character_image.source = str(int(character_image.im_num)) + ".png"
```

```python
def change_monst_im(self):
    monster_image = self.root.ids['monster_image']
    monster_image.source = str(int(monster_image.im_num)) + ".png"

def start_char_animation(self, touch_pos):
    character_image = self.root.ids['character_image']
    character_image.im_num = 0
    char_anim = kivy.animation.Animation(pos_hint={'x': touch_pos[0] -
    character_image.size_hint[0] / 2, 'y': touch_pos[1] - character_
    image.size_hint[1] / 2}, im_num=7)
    char_anim.bind(on_complete=self.char_animation_completed)
    char_anim.start(character_image)

def start_monst_animation(self, new_pos, anim_duration):
    monster_image = self.root.ids['monster_image']
    monst_anim = kivy.animation.Animation(pos_hint={'x': new_pos[0],
    'y': new_pos[1]}, im_num=17, duration=anim_duration)
    monst_anim.bind(on_complete=self.monst_animation_completed)
    monst_anim.start(monster_image)

def build(self):
    coin_width = 0.05
    coin_height = 0.05

    for k in range(TestApp.num_coins):
        x = random.uniform(0, 1 - coin_width)
        y = random.uniform(0, 1 - coin_height)
        coin = kivy.uix.image.Image(source="coin.png", size_hint=(coin_
        width, coin_height), pos_hint={'x': x, 'y': y},allow_stretch=True)
        self.root.add_widget(coin, index=-1)

def on_start(self):
    monster_image = self.root.ids['monster_image']
    new_pos = (random.uniform(0.0, 1 - monster_image.size_hint[0]),
    random.uniform(0.0, 1 - monster_image.size_hint[1]))
    self.start_monst_animation(new_pos=new_pos, anim_duration=random.
    uniform(1.5, 3.5))
```

```
def monst_pos_hint(self):
    character_image = self.root.ids['character_image']
    monster_image = self.root.ids['monster_image']

    character_center = character_image.center
    monster_center = monster_image.center

    gab_x = character_image.width / 2
    gab_y = character_image.height / 2
    if character_image.collide_widget(monster_image) and abs(character_
    center[0] - monster_center[0]) <= gab_x and abs(character_
    center[1] - monster_center[1]) <= gab_y:
        TestApp.character_killed = True

        kivy.animation.Animation.cancel_all(character_image)
        kivy.animation.Animation.cancel_all(monster_image)

        character_image.im_num = 91
        char_anim = kivy.animation.Animation(im_num=95)
        char_anim.start(character_image)

app = TestApp()
app.run()
```

Because the coin's positioning is random, there is a chance that most or even all of them will be in a small area, as shown in Figure 5-24, after running the application with the modified build() function. We need to guarantee that each coin is a minimum distance away from the next coin. The distance could be either horizontally or vertically.

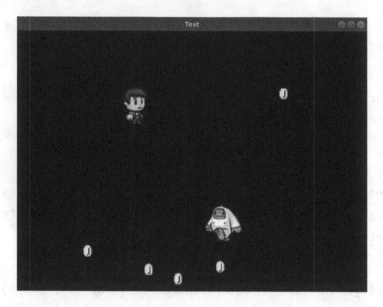

Figure 5-24. *The coins may be very close to each other*

The way to position the coins is to divide the screen into a number of vertical sections equal to the number of coins to be added. This is illustrated in Figure 5-25. One coin is randomly positioned at any position in just one section.

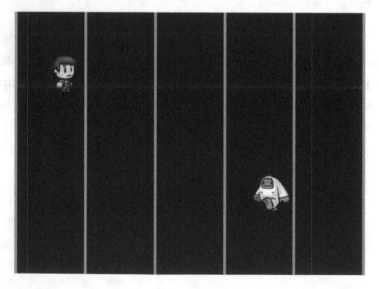

Figure 5-25. *Splitting the screen width equally for placing the coins*

The modified build() function is shown in Listing 5-38. Because the screen is split vertically, each section will cover the entire height of the window but its width is limited according to the number of used coins. For that reason, the section width is calculated in the section_width variable.

Listing 5-38. Splitting the Screen Width Uniformly to Add Multiple Image Widgets Representing the Coins

```
import kivy.app
import kivy.animation
import kivy.uix.image
import random

class TestApp(kivy.app.App):
    character_killed = False
    num_coins = 5
    coins_ids = {}

    def char_animation_completed(self, *args):
        character_image = self.root.ids['character_image']
        character_image.im_num = 0

    def monst_animation_completed(self, *args):
        monster_image = self.root.ids['monster_image']
        monster_image.im_num = 10

        new_pos = (random.uniform(0.0, 1 - monster_image.size_hint[0]),
        random.uniform(0.0, 1 - monster_image.size_hint[1]))
        self.start_monst_animation(new_pos= new_pos,anim_duration=random.
        uniform(1.5, 3.5))

    def touch_down_handler(self, args):
        if TestApp.character_killed == False:
            self.start_char_animation(args[1].spos)

    def change_char_im(self):
        character_image = self.root.ids['character_image']
        character_image.source = str(int(character_image.im_num)) + ".png"
```

```python
    def change_monst_im(self):
        monster_image = self.root.ids['monster_image']
        monster_image.source = str(int(monster_image.im_num)) + ".png"

    def start_char_animation(self, touch_pos):
        character_image = self.root.ids['character_image']
        character_image.im_num = 0
        char_anim = kivy.animation.Animation(pos_hint={'x': touch_pos[0] -
        character_image.size_hint[0] / 2, 'y': touch_pos[1] - character_
        image.size_hint[1] / 2}, im_num=7)
        char_anim.bind(on_complete=self.char_animation_completed)
        char_anim.start(character_image)

    def start_monst_animation(self, new_pos, anim_duration):
        monster_image = self.root.ids['monster_image']
        monst_anim = kivy.animation.Animation(pos_hint={'x': new_pos[0],
        'y': new_pos[1]}, im_num=17, duration=anim_duration)
        monst_anim.bind(on_complete=self.monst_animation_completed)
        monst_anim.start(monster_image)

    def build(self):
        coin_width = 0.05
        coin_height = 0.05

        section_width = 1.0/TestApp.num_coins
        for k in range(TestApp.num_coins):
            x = random.uniform(section_width*k, section_width*(k+1)-coin_
            width)
            y = random.uniform(0, 1-coin_height)
            coin = kivy.uix.image.Image(source="coin.png", size_hint=(coin_
            width, coin_height), pos_hint={'x': x, 'y': y}, allow_
            stretch=True, id='coin'+str(k))
            self.root.add_widget(coin, index=-1)
            TestApp.coins_ids['coin'+str(k)] = coin
```

```python
    def on_start(self):
        monster_image = self.root.ids['monster_image']
        new_pos = (random.uniform(0.0, 1 - monster_image.size_hint[0]),
        random.uniform(0.0, 1 - monster_image.size_hint[1]))
        self.start_monst_animation(new_pos=new_pos, anim_duration=random.
        uniform(1.5, 3.5))

    def monst_pos_hint(self):
        character_image = self.root.ids['character_image']
        monster_image = self.root.ids['monster_image']

        character_center = character_image.center
        monster_center = monster_image.center

        gab_x = character_image.width / 2
        gab_y = character_image.height / 2
        if character_image.collide_widget(monster_image) and abs(character_
        center[0] - monster_center[0]) <= gab_x and abs(character_
        center[1] - monster_center[1]) <= gab_y:
            TestApp.character_killed = True

            kivy.animation.Animation.cancel_all(character_image)
            kivy.animation.Animation.cancel_all(monster_image)

            character_image.im_num = 91
            char_anim = kivy.animation.Animation(im_num=95)
            char_anim.start(character_image)

app = TestApp()
app.run()
```

Each coin can be placed within the boundary of the section. Because there is no restriction on the section height, the coin y coordinate is calculated as previously shown. In order to place the coin within the width specified by the section, the range from which the x coordinate is selected is bounded to its start and end columns. The start value is defined by section_width*k while the end is section_width*(k+1)-coin_width. Note that the coin_width is subtracted to make sure the coin is within the section boundary.

For the first coin, the loop variable k value is 0 and thus the start value is 0.0 but the end value is `section_width-coin_width`. Given that `section_width` equals 0.2 and `coin_width` equals 0.05, the range of the first section is **0.0:0.15**. For the second coin, k will be 1 and thus the start value is `section_width`, while the end value is `section_width*2-coin_width`. Thus, the range of the second section is **0.2:0.35**. The same way, the ranges for the remaining sections are **0.4:0.55**, **0.6:0.75**, and **0.8:0.95**.

We used to use the `ids` dictionary of the root widget to reference a child widget inside the Python file. Unfortunately, the `ids` dictionary does not hold references to the dynamically added widgets inside the Python file. In order to be able to reference such widgets later, their references are saved inside the `coins_ids` dictionary defined in the class header. Each coin is given a string key in that dictionary consisting of the word `coin` appended to the coin number starting with 0. Thus, the keys are `coin0`, `coin1`, `coin2`, `coin3`, and `coin4`.

Figure 5-26 shows the result after running the application. The coins are better distributed. After placing the coins, the next step is to allow the player to collect them.

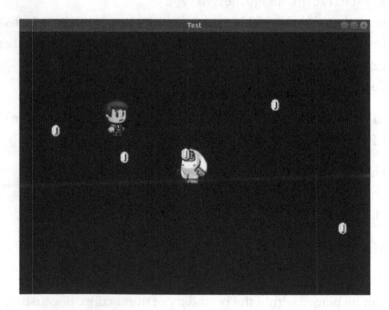

Figure 5-26. *Uniformly distributing the coins across the screen*

The coins positions are printed according to this output:

```
{'coin0': <kivy.uix.image.Image object at 0x7f0c56ff4388>, 'coin1': <kivy.
uix.image.Image object at 0x7f0c56ff44c0>, 'coin2': <kivy.uix.image.
Image object at 0x7f0c56ff4590>, 'coin3': <kivy.uix.image.Image object at
0x7f0c56ff4660>, 'coin4': <kivy.uix.image.Image object at 0x7f0c56ff4730>}
```

Collecting Coins

In order to collect the coins, we need to detect collisions between the character and all the coins not collected yet. In order to access the character position each time it is changed, the on_pos_hint event is bound to the character Image widget inside the KV file. Thus, the modified KV file is listed in Listing 5-39. The event is given the callback function char_pos_hint().

Listing 5-39. Adding the on_pos_hint Event to Return the Character Position

```
FloatLayout:
    on_touch_down: app.touch_down_handler(args)

    Image:
        id: monster_image
        size_hint: (0.15, 0.15)
        pos_hint: {'x': 0.8, 'y': 0.8}
        source: "10.png"
        im_num: 10
        allow_stretch: True
        on_im_num: app.change_monst_im()
        on_pos_hint: app.monst_pos_hint()

    Image:
        id: character_image
        size_hint: (0.15, 0.15)
        pos_hint: {'x': 0.2, 'y': 0.6}
        source: "0.png"
        im_num: 0
        allow_stretch: True
        on_im_num: app.change_char_im()
        on_pos_hint: app.char_pos_hint()
```

According to the implementation of this function in the Python file listed in Listing 5-40, it loops through the items (i.e. coins) in the dictionary and returns the key and value of each item (i.e., coin) in the coin_key and curr_coin variables defined in the loop header. The collision will be detected the same way the collision was detected between the character and the monster.

The `collide_widget()` returns True if there is intersection between the boundaries of the two widgets even in a single row or column. In order to tune it, the centers of the two widgets are compared. If the difference between the centers exceeds a predefined thresholds, it indicates occurrence of a collision.

Once collision occurs between the character and a coin, the coin Image widget is removed from the widget tree by calling the `remove_widget()` method. This ensures that the widget becomes hidden after being collected. Character collision with the coins is detected similar to calculating it with the monster, except for reducing the gab_x and gab_y variables when working with the coins, because their size is smaller than the monster size.

Listing 5-40. Handling the char_pos_hint() Function to Detect Collision with the Coins

```python
import kivy.app
import kivy.animation
import kivy.uix.image
import random

class TestApp(kivy.app.App):
    character_killed = False
    num_coins = 5
    coins_ids = {}

    def char_animation_completed(self, *args):
        character_image = self.root.ids['character_image']
        character_image.im_num = 0

    def monst_animation_completed(self, *args):
        monster_image = self.root.ids['monster_image']
        monster_image.im_num = 10

        new_pos = (random.uniform(0.0, 1 - monster_image.size_hint[0]),
        random.uniform(0.0, 1 - monster_image.size_hint[1]))
        self.start_monst_animation(new_pos= new_pos, anim_duration=random.
        uniform(1.5, 3.5))
```

```python
    def touch_down_handler(self, args):
        if TestApp.character_killed == False:
            self.start_char_animation(args[1].spos)

    def change_char_im(self):
        character_image = self.root.ids['character_image']
        character_image.source = str(int(character_image.im_num)) + ".png"

    def change_monst_im(self):
        monster_image = self.root.ids['monster_image']
        monster_image.source = str(int(monster_image.im_num)) + ".png"

    def start_char_animation(self, touch_pos):
        character_image = self.root.ids['character_image']
        character_image.im_num = 0
        char_anim = kivy.animation.Animation(pos_hint={'x': touch_pos[0] -
        character_image.size_hint[0] / 2, 'y': touch_pos[1] - character_
        image.size_hint[1] / 2}, im_num=7)
        char_anim.bind(on_complete=self.char_animation_completed)
        char_anim.start(character_image)

    def start_monst_animation(self, new_pos, anim_duration):
        monster_image = self.root.ids['monster_image']
        monst_anim = kivy.animation.Animation(pos_hint={'x': new_pos[0],
        'y': new_pos[1]}, im_num=17, duration=anim_duration)
        monst_anim.bind(on_complete=self.monst_animation_completed)
        monst_anim.start(monster_image)

    def build(self):
        coin_width = 0.05
        coin_height = 0.05

        section_width = 1.0/TestApp.num_coins
        for k in range(TestApp.num_coins):
            x = random.uniform(section_width*k, section_width*(k+1)-coin_
            width)
            y = random.uniform(0, 1-coin_height)
```

```python
        coin = kivy.uix.image.Image(source="coin.png", size_hint=(coin_
        width, coin_height), pos_hint={'x': x, 'y': y}, allow_
        stretch=True)
        self.root.add_widget(coin, index=-1)
        TestApp.coins_ids['coin'+str(k)] = coin

    def on_start(self):
        monster_image = self.root.ids['monster_image']
        new_pos = (random.uniform(0.0, 1 - monster_image.size_hint[0]),
        random.uniform(0.0, 1 - monster_image.size_hint[1]))
        self.start_monst_animation(new_pos=new_pos, anim_duration=random.
        uniform(1.5, 3.5))

    def monst_pos_hint(self):
        character_image = self.root.ids['character_image']
        monster_image = self.root.ids['monster_image']

        character_center = character_image.center
        monster_center = monster_image.center

        gab_x = character_image.width / 2
        gab_y = character_image.height / 2
        if character_image.collide_widget(monster_image) and abs(character_
        center[0] - monster_center[0]) <= gab_x and abs(character_
        center[1] - monster_center[1]) <= gab_y:
            TestApp.character_killed = True

            kivy.animation.Animation.cancel_all(character_image)
            kivy.animation.Animation.cancel_all(monster_image)

            character_image.im_num = 91
            char_anim = kivy.animation.Animation(im_num=95)
            char_anim.start(character_image)

    def char_pos_hint(self):
        character_image = self.root.ids['character_image']
        character_center = character_image.center

        gab_x = character_image.width / 3
        gab_y = character_image.height / 3
```

```
    for coin_key, curr_coin in TestApp.coins_ids.items():
        curr_coin_center = curr_coin.center
        if character_image.collide_widget(curr_coin) and abs(character_
        center[0] - curr_coin_center[0]) <= gab_x and abs(character_
        center[1] - curr_coin_center[1]) <= gab_y:
            print("Coin Collected", coin_key)
            self.root.remove_widget(curr_coin)
```

```
app = TestApp()
app.run()
```

There is an issue in the previous application. Even after the coin is deleted from the widget tree, there is still an item for it in the dictionary. As a result, the loop goes though five iterations even if all of the items are collected and behaves as if no coins were not collected.

In order to make sure the widget is detected from the dictionary, we can keep track of the coins collected in an empty list named coins_to_delete defined inside the char_pos_hint() function, as listed in Listing 5-41. For each coin collected, its key in the coins_ids dictionary is appended to the list using the append() function.

Listing 5-41. Removing the Coins Once They Are Collected

```
def char_pos_hint(self):
    character_image = self.root.ids['character_image']
    character_center = character_image.center

    gab_x = character_image.width / 3
    gab_y = character_image.height / 3
    coins_to_delete = []

    for coin_key, curr_coin in TestApp.coins_ids.items():
        curr_coin_center = curr_coin.center
        if character_image.collide_widget(curr_coin) and abs(character_
        center[0] - curr_coin_center[0]) <= gab_x and abs(character_
        center[1] - curr_coin_center[1]) <= gab_y:
            print("Coin Collected", coin_key)
            coins_to_delete.append(coin_key)
            self.root.remove_widget(curr_coin)
```

```
if len(coins_to_delete) > 0:
    for coin_key in coins_to_delete:
        del TestApp.coins_ids[coin_key]
```

After the loop ends, an `if` statement determines whether the list is empty or not based on its length. If its length is less than 1, there are no coins collected in the previous loop and thus no items (i.e. coins) to delete from the dictionary. If the length of the list is higher than or equal to 1 (i.e. higher than 0), this means that there are some coins from the previous loop.

In order to delete the coins from the dictionary, a `for` loop iterates through the elements in the list. Note that the list elements represent the keys of each coin, such as `coin0`. Thus, the key stored in the list will be used as an index to the dictionary to return the associated coin Image widget. Using the `del` command in Python, the item can be deleted from the dictionary. By doing that, we have completely deleted the coin from both the widget tree and the dictionary. After collecting all coins, the number of items in the dictionary will be zero and the loop will be useless.

Complete Level

In the previous application, there was no indication about the number of collected coins. According to the modified KV file shown in Listing 5-42, a small `Label` widget is added to the top-left corner of the screen to display the number of collected coins. The label is given an ID of `num_coins_collected` in order to change its text inside the Python code.

Listing 5-42. Displaying the Number of Collected Coins in a Label Widget Placed at the Top of the Screen

```
FloatLayout:
    on_touch_down: app.touch_down_handler(args)

    Label:
        id: num_coins_collected
        size_hint: (0.1, 0.02)
        pos_hint: {'x': 0.0, 'y': 0.97}
        text: "Coins 0"
        font_size: 20
```

```
Image:
    id: monster_image
    size_hint: (0.15, 0.15)
    pos_hint: {'x': 0.8, 'y': 0.8}
    source: "10.png"
    im_num: 10
    allow_stretch: True
    on_im_num: app.change_monst_im()
    on_pos_hint: app.monst_pos_hint()

Image:
    id: character_image
    size_hint: (0.15, 0.15)
    pos_hint: {'x': 0.2, 'y': 0.6}
    source: "0.png"
    im_num: 0
    allow_stretch: True
    on_im_num: app.change_char_im()
    on_pos_hint: app.char_pos_hint()
```

The char_pos_hint() function inside the Python file is changed to update the added label text field according to the number of currently collected coins. The file is shown in Listing 5-43. First, a variable named num_coins_collected is defined in the class and given an initial value of 0. If a collision occurred between the character and any coin, then that variable gets incremented by 1 and then the Label widget gets updated.

Because completing the mission of collecting all coins means the end of the current level, it is better to do something to indicate the end of the level. If the number of collected coins inside the num_coins_collected variable is equal to the number of coins in the num_coins variable, a label is dynamically added in the widget tree and it displays the "Level Completed" message. Besides creating this widget, the character and monster animations are cancelled. Note that by cancelling the monster animation, its position will not change and thus the monst_pos_hint() callback function will not be executed.

Listing 5-43. Updating the Label Displaying the Number of Collected Coins and Displaying a Message When the Level Completes

```python
import kivy.app
import kivy.animation
import kivy.uix.image
import kivy.uix.label
import random

class TestApp(kivy.app.App):
    character_killed = False
    num_coins = 5
    num_coins_collected = 0
    coins_ids = {}

    def char_animation_completed(self, *args):
        character_image = self.root.ids['character_image']
        character_image.im_num = 0

    def monst_animation_completed(self, *args):
        monster_image = self.root.ids['monster_image']
        monster_image.im_num = 10

        new_pos = (random.uniform(0.0, 1 - monster_image.size_hint[0]),
        random.uniform(0.0, 1 - monster_image.size_hint[1]))
        self.start_monst_animation(new_pos= new_pos,anim_duration=random.
        uniform(1.5, 3.5))

    def touch_down_handler(self, args):
        if TestApp.character_killed == False:
            self.start_char_animation(args[1].spos)

    def change_char_im(self):
        character_image = self.root.ids['character_image']
        character_image.source = str(int(character_image.im_num)) + ".png"

    def change_monst_im(self):
        monster_image = self.root.ids['monster_image']
        monster_image.source = str(int(monster_image.im_num)) + ".png"
```

```python
def start_char_animation(self, touch_pos):
    character_image = self.root.ids['character_image']
    character_image.im_num = 0
    char_anim = kivy.animation.Animation(pos_hint={'x': touch_pos[0] -
    character_image.size_hint[0] / 2, 'y': touch_pos[1] - character_
    image.size_hint[1] / 2}, im_num=7)
    char_anim.bind(on_complete=self.char_animation_completed)
    char_anim.start(character_image)

def start_monst_animation(self, new_pos, anim_duration):
    monster_image = self.root.ids['monster_image']
    monst_anim = kivy.animation.Animation(pos_hint={'x': new_pos[0],
    'y': new_pos[1]}, im_num=17, duration=anim_duration)
    monst_anim.bind(on_complete=self.monst_animation_completed)
    monst_anim.start(monster_image)

def build(self):
    coin_width = 0.05
    coin_height = 0.05

    section_width = 1.0/TestApp.num_coins
    for k in range(TestApp.num_coins):
        x = random.uniform(section_width*k, section_width*(k+1)-coin_
            width)
        y = random.uniform(0, 1-coin_height)
        coin = kivy.uix.image.Image(source="coin.png", size_hint=(coin_
        width, coin_height), pos_hint={'x': x, 'y': y}, allow_
        stretch=True)
        self.root.add_widget(coin, index=-1)
        TestApp.coins_ids['coin'+str(k)] = coin

def on_start(self):
    monster_image = self.root.ids['monster_image']
    new_pos = (random.uniform(0.0, 1 - monster_image.size_hint[0]),
    random.uniform(0.0, 1 - monster_image.size_hint[1]))
    self.start_monst_animation(new_pos=new_pos, anim_duration=random.
    uniform(1.5, 3.5))
```

```
    def monst_pos_hint(self):
        character_image = self.root.ids['character_image']
        monster_image = self.root.ids['monster_image']

        character_center = character_image.center
        monster_center = monster_image.center

        gab_x = character_image.width / 2
        gab_y = character_image.height / 2
        if character_image.collide_widget(monster_image) and abs(character_
        center[0] - monster_center[0]) <= gab_x and abs(character_
        center[1] - monster_center[1]) <= gab_y:
            TestApp.character_killed = True

            kivy.animation.Animation.cancel_all(character_image)
            kivy.animation.Animation.cancel_all(monster_image)

            character_image.im_num = 91
            char_anim = kivy.animation.Animation(im_num=95)
            char_anim.start(character_image)

    def char_pos_hint(self):
        character_image = self.root.ids['character_image']
        character_center = character_image.center

        gab_x = character_image.width / 3
        gab_y = character_image.height / 3
        coins_to_delete = []

        for coin_key, curr_coin in TestApp.coins_ids.items():
            curr_coin_center = curr_coin.center
            if character_image.collide_widget(curr_coin) and abs(character_
            center[0] - curr_coin_center[0]) <= gab_x and abs(character_
            center[1] - curr_coin_center[1]) <= gab_y:
                coins_to_delete.append(coin_key)
                self.root.remove_widget(curr_coin)
                TestApp.num_coins_collected = TestApp.num_coins_collected + 1
                self.root.ids['num_coins_collected'].text = "Coins
                "+str(TestApp.num_coins_collected)
```

```
        if TestApp.num_coins_collected == TestApp.num_coins:
            kivy.animation.Animation.cancel_all(character_image)
            kivy.animation.Animation.cancel_all(self.root.
            ids['monster_image'])
            self.root.add_widget(kivy.uix.label.Label(pos_
            hint={'x': 0.1, 'y': 0.1}, size_hint=(0.8, 0.8), font_
            size=90, text="Level Completed"))

    if len(coins_to_delete) > 0:
        for coin_key in coins_to_delete:
            del TestApp.coins_ids[coin_key]

app = TestApp()
app.run()
```

Figure 5-27 shows the result after the level is completed.

Figure 5-27. *A message is displayed when the level completes*

Sound Effects

A game without sound effects is not a very good game. Sound is an important factor of the user experience. You can add sound effects to every action happening in the game. For our game, we will add sound effects to the character's death, when completing a level, and when collecting coins. This is in addition to the background music, which helps the players engage with the game.

Kivy provides a very simple interface for playing sounds using the SoundLoader class found in the kivy.core.audio module. The modified Python file in which sounds are loaded and played is shown in Listing 5-44.

Listing 5-44. Adding Sound Effects to the Game

```python
import kivy.app
import kivy.animation
import kivy.uix.image
import kivy.uix.label
import random
import kivy.core.audio
import os

class TestApp(kivy.app.App):
    character_killed = False
    num_coins = 5
    num_coins_collected = 0
    coins_ids = {}

    def char_animation_completed(self, *args):
        character_image = self.root.ids['character_image']
        character_image.im_num = 0

    def monst_animation_completed(self, *args):
        monster_image = self.root.ids['monster_image']
        monster_image.im_num = 10

        new_pos = (random.uniform(0.0, 1 - monster_image.size_hint[0]),
        random.uniform(0.0, 1 - monster_image.size_hint[1]))
        self.start_monst_animation(new_pos= new_pos,anim_duration=random.
        uniform(1.5, 3.5))
```

```python
    def touch_down_handler(self, args):
        if TestApp.character_killed == False:
            self.start_char_animation(args[1].spos)

    def change_char_im(self):
        character_image = self.root.ids['character_image']
        character_image.source = str(int(character_image.im_num)) + ".png"

    def change_monst_im(self):
        monster_image = self.root.ids['monster_image']
        monster_image.source = str(int(monster_image.im_num)) + ".png"

    def start_char_animation(self, touch_pos):
        character_image = self.root.ids['character_image']
        character_image.im_num = 0
        char_anim = kivy.animation.Animation(pos_hint={'x': touch_pos[0] -
        character_image.size_hint[0] / 2,'y': touch_pos[1] - character_
        image.size_hint[1] / 2}, im_num=7)
        char_anim.bind(on_complete=self.char_animation_completed)
        char_anim.start(character_image)

    def start_monst_animation(self, new_pos, anim_duration):
        monster_image = self.root.ids['monster_image']
        monst_anim = kivy.animation.Animation(pos_hint={'x': new_pos[0],
'y': new_pos[1]}, im_num=17,duration=anim_duration)
        monst_anim.bind(on_complete=self.monst_animation_completed)
        monst_anim.start(monster_image)

    def build(self):
        coin_width = 0.05
        coin_height = 0.05

        section_width = 1.0/TestApp.num_coins
        for k in range(TestApp.num_coins):
            x = random.uniform(section_width*k, section_width*(k+1)-coin_
                width)
            y = random.uniform(0, 1-coin_height)
```

```
            coin = kivy.uix.image.Image(source="coin.png", size_hint=(coin_
            width, coin_height), pos_hint={'x': x, 'y': y}, allow_
            stretch=True)
            self.root.add_widget(coin, index=-1)
            TestApp.coins_ids['coin'+str(k)] = coin

    def on_start(self):
        music_dir = os.getcwd()+"/music/"
        self.bg_music = kivy.core.audio.SoundLoader.load(music_dir+
        "bg_music_piano.wav")
        self.bg_music.loop = True

        self.coin_sound = kivy.core.audio.SoundLoader.load(music_dir+
        "coin.wav")
        self.level_completed_sound = kivy.core.audio.SoundLoader.
        load(music_dir+"level_completed_flaute.wav")
        self.char_death_sound = kivy.core.audio.SoundLoader.
        load(music_dir+"char_death_flaute.wav")

        self.bg_music.play()

        monster_image = self.root.ids['monster_image']
        new_pos = (random.uniform(0.0, 1 - monster_image.size_hint[0]),
        random.uniform(0.0, 1 - monster_image.size_hint[1]))
        self.start_monst_animation(new_pos=new_pos, anim_duration=random.
        uniform(1.5, 3.5))

    def monst_pos_hint(self):
        character_image = self.root.ids['character_image']
        monster_image = self.root.ids['monster_image']

        character_center = character_image.center
        monster_center = monster_image.center

        gab_x = character_image.width / 2
        gab_y = character_image.height / 2
        if character_image.collide_widget(monster_image) and abs(character_
        center[0] - monster_center[0]) <= gab_x and abs(character_
        center[1] - monster_center[1]) <= gab_y:
```

```python
        self.bg_music.stop()
        self.char_death_sound.play()
        TestApp.character_killed = True

        kivy.animation.Animation.cancel_all(character_image)
        kivy.animation.Animation.cancel_all(monster_image)

        character_image.im_num = 91
        char_anim = kivy.animation.Animation(im_num=95)
        char_anim.start(character_image)

def char_pos_hint(self):
    character_image = self.root.ids['character_image']
    character_center = character_image.center

    gab_x = character_image.width / 3
    gab_y = character_image.height / 3
    coins_to_delete = []

    for coin_key, curr_coin in TestApp.coins_ids.items():
        curr_coin_center = curr_coin.center
        if character_image.collide_widget(curr_coin) and abs(character_
        center[0] - curr_coin_center[0]) <= gab_x and abs(character_
        center[1] - curr_coin_center[1]) <= gab_y:
            self.coin_sound.play()
            coins_to_delete.append(coin_key)
            self.root.remove_widget(curr_coin)
            TestApp.num_coins_collected = TestApp.num_coins_collected + 1
            self.root.ids['num_coins_collected'].text = "Coins
            "+str(TestApp.num_coins_collected)
            if TestApp.num_coins_collected == TestApp.num_coins:
                self.bg_music.stop()
                self.level_completed_sound.play()
                kivy.animation.Animation.cancel_all(character_image)
                kivy.animation.Animation.cancel_all(self.root.
                ids['monster_image'])
```

```
                    self.root.add_widget(kivy.uix.label.Label(pos_
                    hint={'x': 0.1, 'y': 0.1}, size_hint=(0.8, 0.8), font_
                    size=90, text="Level Completed"))

        if len(coins_to_delete) > 0:
            for coin_key in coins_to_delete:
                del TestApp.coins_ids[coin_key]
app = TestApp()
app.run()
```

There are two steps to playing a sound file. We must first load the sound file using the `load()` method of the `SoundLoader` class. This method accepts the sound file path that is specified in the `music_dir` variable. This variable uses the `os` module to return the current directory using the `os.getcwd()` function. Assuming that the sound files are stored in a folder named `music` inside the current directory, the complete path for the files is the concatenation between `os.getcwd()` and the file called `music`.

All the sound files are prepared in the `on_start()` method of the `TestApp` class of the application. The background sound file is loaded into the `bg_music` variable. The sound files for collecting coins, character death, and level completion are stored into the variables `coin_sound`, `char_death_sound`, and `level_completed_sound`, respectively. Note that each of these variables is associated with `self`, which refers to the current object. This helps to control the sound files even outside the `on_start()` method. Remember to use `self` when referencing the sound files outside that method.

The second step is to play the file using the `play()` method. For the background music, it is played within the `on_start()` method. The coin sound is played after a collision occurs with a coin inside the `char_pos_hint()` callback function.

The level completion sound is played after collecting all the coins. Because the level is completed, there is no need for the background music anymore and thus it is stopped by calling the `stop()` method.

Finally, the character death sound is played inside the `monst_pos_hint()` callback function after collision occurs with the monster. Playing the game after adding sound effects is more interesting than before.

Game Background

We can change the background of the game to be more attractive rather than the default black background. You can use a texture, an animated image, or a static image as the background.

A static image is used as the background of the game according to the KV file shown in Listing 5-45. It is drawn inside the FloatLayout using canvas.before. This guarantees that the image will cover the entire window.

Listing 5-45. Adding a Background Image to the Game

```
FloatLayout:
    on_touch_down: app.touch_down_handler(args)
    canvas.before:
        Rectangle:
            size: self.size
            pos: self.pos
            source: 'bg.jpg'

    Label:
        id: num_coins_collected
        size_hint: (0.1, 0.02)
        pos_hint: {'x': 0.0, 'y': 0.97}
        text: "Coins 0"
        font_size: 20

    Image:
        id: monster_image
        size_hint: (0.15, 0.15)
        pos_hint: {'x': 0.8, 'y': 0.8}
        source: "10.png"
        im_num: 10
        allow_stretch: True
        on_im_num: app.change_monst_im()
        on_pos_hint: app.monst_pos_hint()
```

```
Image:
    id: character_image
    size_hint: (0.15, 0.15)
    pos_hint: {'x': 0.2, 'y': 0.6}
    source: "0.png"
    im_num: 0
    allow_stretch: True
    on_im_num: app.change_char_im()
    on_pos_hint: app.char_pos_hint()
```

Figure 5-28 shows the game after adding the background.

Figure 5-28. *Adding a background image to the screen*

Game Development Recap

The game currently has just one level and we need to add more levels to it. Before adding more levels, it is important to take a general overview of the progress in the game development up to this point.

Figure 5-29 shows the flow of game execution. Because our Kivy application implements the build() and on_start() methods, according to the Kivy application lifecycle, they will be executed before any of our custom functions. This starts at the build() function until ending the game because the character was killed or all the coins were collected and the level is complete. Each function is listed in order to its execution until reaching an end for the game and its tasks are listed to its right.

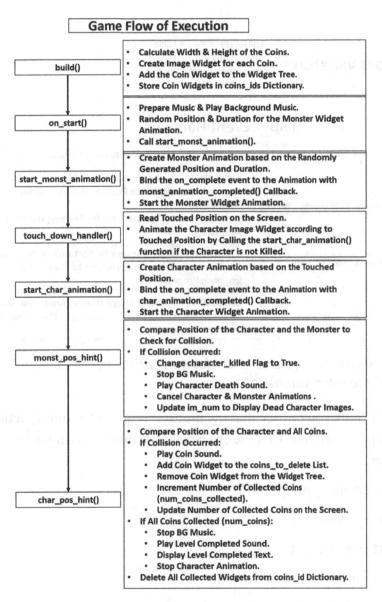

Figure 5-29. *The flow of game execution*

Figure 5-30 lists the callback functions that handle events that serve the animation of the character and monster. It also lists the five class variables used in the previous application.

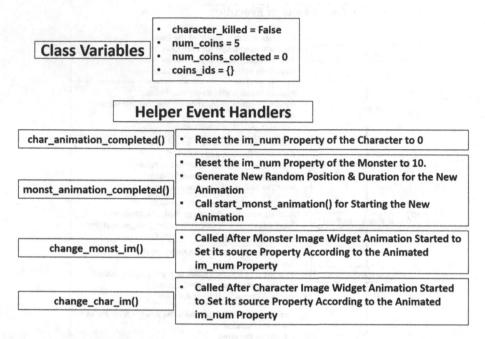

Figure 5-30. *Summary of class variables and callback functions to handle the character and monster animations*

From Figure 5-30, you can see that the monster requires the four functions listed as follows for its operation. Note that the character uses functions similar to these, but with a change in their names.

- start_monst_animation()

- change_monst_im()

- monst_pos_hint()

- monst_animation_completed()

Summary

The game now has a character that moves, using animation, according to the touch position. A monster moves randomly, also using animation. The character is killed when it collides with the monster. When it's killed, a one-time animation starts that changes the image of the character to reflect death. A number of coins are uniformly distributed on the screen where the player's mission is to collect all of them. A label at the top of the screen shows the number of collected coins. When collision occurs between the character and a coin, the coin disappears and the label is updated. When all the coins are collected, the level is complete.

## CHAPTER 6

Completing and Publishing Your First Game

In the previous chapter, we started developing a multi-level cross-platform game in which a character moves, using animation, according to the touch position. The player's job is to collect a number of uniformly distributed coins on the screen. A monster tries to attack the player in order to kill it. When a collision occurs between the player and the monster, the player dies and the level ends. Until the end of the previous chapter, the game just had a single level.

In this chapter, we continue developing the game by adding more levels. The screens introduced in Chapter 4 will be used to organize the game interface. More monsters will be added. Fires will be introduced that go back and forth through a predefined path. As we develop the game, important issues will be discussed and resolved. At the end of the next chapter, the game will be published for Android devices on Google Play for any user to download and install.

Adding More Levels to the Game

Our game has a single level in which the player has to collect five coins to complete it. After the mission is completed, the "Level Completed" message is displayed in a Label widget. In this section, more levels will be added to the game so that the player can go from one level to another. When building this multi-level game, we will try to follow the don't repeat yourself (DRY) principle as much as possible.

© Ahmed Fawzy Mohamed Gad 2019
A. F. M. Gad, *Building Android Apps in Python Using Kivy with Android Studio*,
https://doi.org/10.1007/978-1-4842-5031-0_6

The best way to create a multi-level game in Kivy is to use multiple screens where each screen holds the widget tree of one level. As a result, we are going to use the Screen and the ScreenManager classes to build and manage multiple screens. Let's start by creating the template we will use when developing the game.

Creating Multiple Screens

As we discussed in Chapter 4, a class is created inside the Python file for each screen to create. So, in order to create an application with two screens, three classes are created in the Python file. One class for the application and two classes for the two screens. In our game, we will start adding another level beside the previous level. So, there is a total of two levels. Each level will have a screen and thus a new class is created for each screen. There will be an additional screen to be used as the game main screen from which the player can go to any level. As a result, the Python file shown in Listing 6-1 will have three screens. The classes created for the two screens representing the levels are named Level1 and Level2. The main screen class is named MainScreen. Because the classes are empty, the pass keyword is added as a virtual class body.

Listing 6-1. Adding a Level and a Main Screen to the Game Using Screens

```python
import kivy.app
import kivy.uix.screenmanager

class TestApp(kivy.app.App):
    pass

class MainScreen(kivy.uix.screenmanager.Screen):
    pass

class Level1(kivy.uix.screenmanager.Screen):
    pass

class Level2(kivy.uix.screenmanager.Screen):
    pass

app = TestApp()
app.run()
```

Inside the KV file in Listing 6-2, the `ScreenManager` is used as the root widget in the application widget tree. It has three children, which are the instances from the three screens classes with names `level1` for the first screen, `level2` for the second, and `main` for the main screen. It is possible to define the layout of each screen in the KV file by enclosing the screen class name between `<>`.

Remember that the level we previously created has a `FloatLayout` for holding all game widgets. Because each screen represents a game level, it will have a child `FloatLayout` to hold all widgets inside the level. This layout will have a background image associated using the `Rectangle` vertex instruction inside `canvas.before`. The name of the images are `bg_lvl1.jpg` and `bg_lvl2.jpg` for the first and second levels, respectively.

Listing 6-2. Defining the Widget Tree of the Second Level of the Game

```
ScreenManager:
    MainScreen:
    Level1:
    Level2:

<MainScreen>:
    name: "main"
    BoxLayout:
        Button:
            text: "Go to Level 1"
            on_press: app.root.current="level1"
        Button:
            text: "Go to Level 2"
            on_press: app.root.current="level2"
<Level1>:
    name: "level1"
    FloatLayout:
        canvas.before:
            Rectangle:
                size: self.size
                pos: self.pos
                source: "bg_lvl1.jpg"
```

```
<Level2>:
    name: "level2"
    FloatLayout:
        canvas.before:
            Rectangle:
                size: self.size
                pos: self.pos
                source: "bg_lvl2.jpg"
```

Remember that the ScreenManager class has a property named current that accepts the name of the screen to be displayed in the window. If it's not explicitly specified, it defaults to the first screen added to the widget tree, which is the MainScreen screen. This screen has a BoxLayout with two child buttons. Each button is responsible for the transition to a game level. After we run the application, we'll see the window in Figure 6-1.

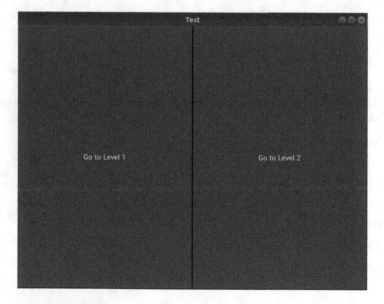

Figure 6-1. *The main screen of the game with two levels*

By preparing these KV and Python files, we set the template to follow for building a game with two levels. Building more levels will be a repetition of the steps we will discuss later.

Adding Custom Widgets in the KV File

In the previous game, the single level had three widgets inside the FloatLayout. The widgets are a Label for printing the number of collected coins, an Image widget for the monster, and another Image widget for the character. To build the two levels, we don't have to duplicate the three widgets for each screen.

Rather than create a separate Image widget for each level, we can create it just once and use it multiple times. This is by creating a custom widget for each element in the game and then reusing it inside each screen. The modified KV file in Listing 6-3 creates the three custom widgets required in each level. At the beginning, there will be minor differences between the two levels. Later, we will add more monsters and coins in the second level.

The two custom widgets named Monster and Character represent the monster and the character, respectively. They are made to extend the Image widget. The extension in KV is made using the @ character. Remember to enclose the name of the custom widgets using <>.

Another custom widget with the name NumCoinsCollected is created and it extends the Label widget to print the number of collected coins. The code inside each widget is identical to what we used in the last game.

Listing 6-3. Creating Custom Widgets for Game Elements

```
ScreenManager:
    MainScreen:
    Level1:
    Level2:

<MainScreen>:
    name: "main"
    BoxLayout:
        Button:
            text: "Go to Level 1"
            on_press: app.root.current="level1"
        Button:
            text: "Go to Level 2"
            on_press: app.root.current="level2"
```

```
<Level1>:
    name: "level1"
    FloatLayout:
        canvas.before:
            Rectangle:
                size: self.size
                pos: self.pos
                source: "bg_lvl1.jpg"
        NumCollectedCoins:
        Monster:
        Character:

<Level2>:
    name: "level2"
    FloatLayout:
        canvas.before:
            Rectangle:
                size: self.size
                pos: self.pos
                source: "bg_lvl2.jpg"
        NumCollectedCoins:
        Monster:
        Character:

<NumCollectedCoins@Label>:
    id: num_coins_collected
    size_hint: (0.1, 0.02)
    pos_hint: {'x': 0.0, 'y': 0.97}
    text: "Coins 0"
    font_size: 20

<Monster@Image>:
    id: monster_image
    size_hint: (0.15, 0.15)
    pos_hint: {'x': 0.8, 'y': 0.8}
    source: "10.png"
    im_num: 10
```

```
    allow_stretch: True
    on_im_num: app.change_monst_im(self)
    on_pos_hint: app.monst_pos_hint(self)
<Character@Image>:
    id: character_image
    size_hint: (0.15, 0.15)
    pos_hint: {'x': 0.2, 'y': 0.6}
    source: "0.png"
    im_num: 0
    allow_stretch: True
    on_im_num: app.change_char_im(self)
    on_pos_hint: app.char_pos_hint(self)
```

Remember that the on_im_num and on_pos_hint events are bound to the Monster and Character widgets. The events for all Character widgets across all levels are handled using the same callback functions, which are change_char_im() and char_pos_hint(), which exist in the application class TestApp. The same process holds for the Monster widget. Because the same function will handle events fired from different widgets, it is important to know which widget fired the event. This is why functions accept the self argument, which refers to the widget that fired the event.

Assigning IDs to Widgets

In order to use these three custom widgets inside each screen, we create an instance. There will be a problem if more than one instance is taken from either custom widget. The custom widgets are given IDs. Because each instance of such widgets inherits all of its properties, all instances will, if not changed, have the same ID.

In Kivy, no two widgets in the same widget tree can have the same ID. Note that the two screens are still within the same widget tree, as they are grouped under the ScreenManager root widget. For this reason, we should remove the ID from the custom widgets templates and add it inside the instances, as shown in the new KV file in Listing 6-4.

For easy manipulation, the IDs for the same widget reflect the index of the screen in which they reside. Moreover, the same widget across the different screens is given the same name except for the screen index at the end of its ID. For example, the Monster widget in the first screen is given an ID monster_image_lvl1, reflecting that it resides in the screen with index 1. This widget is given an ID `monster_image_lvl2` in the screen with index 2. Note that the main screen has index 0.

Because we need to add coins to the `FloatLayout` widget of each screen, they are given IDs to reference them within the Python code. They are given the IDs `layout_lvl1` and `layout_lvl2`.

Listing 6-4. Removing IDs from Custom Widgets and Adding Them Inside the Instances

```
ScreenManager:
    MainScreen:
    Level1:
    Level2:

<MainScreen>:
    name: "main"
    BoxLayout:
        Button:
            text: "Go to Level 1"
            on_press: app.root.current="level1"
        Button:
            text: "Go to Level 2"
            on_press: app.root.current="level2"
<Level1>:
    name: "level1"
    FloatLayout:
        id: layout_lvl1
        on_touch_down: app.touch_down_handler(1, args)
        canvas.before:
            Rectangle:
                size: self.size
                pos: self.pos
                source: "bg_lvl1.jpg"
        NumCollectedCoins:
            id: num_coins_collected_lvl1
```

```
        Monster:
            id: monster_image_lvl1
        Character:
            id: character_image_lvl1
<Level2>:
    name: "level2"
    FloatLayout:
        id: layout_lvl2
        on_touch_down: app.touch_down_handler(2, args)
        canvas.before:
            Rectangle:
                size: self.size
                pos: self.pos
                source: "bg_lvl2.jpg"
        NumCollectedCoins:
            id: num_coins_collected_lvl2
        Monster:
            id: monster_image_lvl2
        Character:
            id: character_image_lvl2

<NumCollectedCoins@Label>:
    size_hint: (0.1, 0.02)
    pos_hint: {'x': 0.0, 'y': 0.97}
    text: "Coins 0"
    font_size: 20

<Monster@Image>:
    size_hint: (0.15, 0.15)
    pos_hint: {'x': 0.8, 'y': 0.8}
    source: "10.png"
    im_num: 10
    allow_stretch: True
    on_im_num: app.change_monst_im(self)
    on_pos_hint: app.monst_pos_hint(self)
```

```
<Character@Image>:
    size_hint: (0.15, 0.15)
    pos_hint: {'x': 0.2, 'y': 0.6}
    source: "0.png"
    im_num: 0
    allow_stretch: True
    on_im_num: app.change_char_im(self)
    on_pos_hint: app.char_pos_hint(self)
```

Note that we need to bind the on_touch_down event to the FloatLayout. This is because the main character in the game moves according to the touched position. From the previous game, remember that we created the callback function touch_down_handler() to handle the event that accepts the args argument.

Note that the same event handler will be used by all screens. Thus, it is very important to know which screen fired the event in order to handle it correctly. This concept holds for all shared functions inside the Python file. Therefore, an additional argument representing the screen index is added to the function. By doing that, we can refer to the exact screen that fired the event.

This is most of the work needed for the KV file to create the two game levels. There will be simple additions later. Let's move on to the Python file.

Game Class Variables

In the previous single level game, there were four variables defined in the TestApp class, which are character_killed, num_coins, num_coins_collected, and coins_ids. Note that if a variable is added to the application class TestApp, it will be shared across each screen. Thus, the same variable will be used by both screens. To solve this issue, such variables will be defined in each screen class, as illustrated in the Python file in Listing 6-5.

Listing 6-5. Adding Variables Inside the Level Classes

```
import kivy.app
import kivy.uix.screenmanager

class TestApp(kivy.app.App):
    pass
```

```python
class MainScreen(kivy.uix.screenmanager.Screen):
    pass

class Level1(kivy.uix.screenmanager.Screen):
    character_killed = False
    num_coins = 5
    num_coins_collected = 0
    coins_ids = {}

class Level2(kivy.uix.screenmanager.Screen):
    character_killed = False
    num_coins = 8
    num_coins_collected = 0
    coins_ids = {}

app = TestApp()
app.run()
```

Screen on_pre_enter Event

According to Figures 5-31 and 5-32 in Chapter 5, which summarize the game flow of execution, the build() method is the first function to be executed in our code according to the Kivy application lifecycle. This method is used to add the coin's widgets to the widget tree.

Because this function is executed in the application level, it will not differentiate between the different screens. Because we need to add coins to each screen, we should use a method that is called per screen. As a result, the build() method is no longer suitable.

The Screen class provides an event called on_pre_enter that is fired when the screen is about to be used before being displayed. Its callback function will be a good place to add the coin's widgets. This event is bound to each screen in the KV file in Listing 6-6. The callback function called screen_on_pre_enter() handles such event for both screens. In order to know which screen fired the event, this callback function accepts an argument referring to the screen index within the ScreenManager.

Listing 6-6. Binding the on_pre_enter Event to Each Screen

```
ScreenManager:
    MainScreen:
    Level1:
    Level2:

<MainScreen>:
    name: "main"
    BoxLayout:
        Button:
            text: "Go to Level 1"
            on_press: app.root.current="level1"
        Button:
            text: "Go to Level 2"
            on_press: app.root.current="level2"

<Level1>:
    name: "level1"
    on_pre_enter: app.screen_on_pre_enter(1)
    FloatLayout:
        id: layout_lvl1
        on_touch_down: app.touch_down_handler(1, args)
        canvas.before:
            Rectangle:
                size: self.size
                pos: self.pos
                source: "bg_lvl1.jpg"
        NumCollectedCoins:
            id: num_coins_collected_lvl1
        Monster:
            id: monster_image_lvl1
        Character:
            id: character_image_lvl1

<Level2>:
    name: "level2"
    on_pre_enter: app.screen_on_pre_enter(2)
```

```
    FloatLayout:
        id: layout_lvl2
        on_touch_down: app.touch_down_handler(2, args)
        canvas.before:
            Rectangle:
                size: self.size
                pos: self.pos
                source: "bg_lvl2.jpg"
        NumCollectedCoins:
            id: num_coins_collected_lvl2
        Monster:
            id: monster_image_lvl2
        Character:
            id: character_image_lvl2
<NumCollectedCoins@Label>:
    size_hint: (0.1, 0.02)
    pos_hint: {'x': 0.0, 'y': 0.97}
    text: "Coins 0"
    font_size: 20

<Monster@Image>:
    size_hint: (0.15, 0.15)
    pos_hint: {'x': 0.8, 'y': 0.8}
    source: "10.png"
    im_num: 10
    allow_stretch: True
    on_im_num: app.change_monst_im(self)
    on_pos_hint: app.monst_pos_hint(self)

<Character@Image>:
    size_hint: (0.15, 0.15)
    pos_hint: {'x': 0.2, 'y': 0.6}
    source: "0.png"
    im_num: 0
    allow_stretch: True
    on_im_num: app.change_char_im(self)
    on_pos_hint: app.char_pos_hint(self)
```

Adding Coins to Game Levels

Inside the screen_on_pre_enter() callback function, the coins are added to the FloatLayout widget of each screen, as shown in the Python file in Listing 6-7. Note how the num_coins screen class variable is retrieved. It is returned using its ID. According to the screen_num argument in the callback function header, the number is appended to the layout_lvl string in order to return the layout inside the screen that fired the on_pre_enter event.

After creating the coin Image widget, according to the randomly generated positions using the uniform() function in the random module, the widget is added inside the FloatLayout widget of the screen.

Finally, the widget reference is inserted into the coins_ids dictionary defined in the screen class. It is accessed in a similar way to the num_coins variable.

Listing 6-7. Adding Coins to the Screen Inside the screen_on_pre_enter() Callback Function

```
import kivy.app
import kivy.uix.screenmanager
import random

class TestApp(kivy.app.App):

    def screen_on_pre_enter(self, screen_num):
        coin_width = 0.05
        coin_height = 0.05

        curr_screen = self.root.screens[screen_num]

        section_width = 1.0/curr_screen.num_coins
        for k in range(curr_screen.num_coins):
            x = random.uniform(section_width*k, section_width*(k+1)-coin_
            width)
            y = random.uniform(0, 1-coin_height)
            coin = kivy.uix.image.Image(source="coin.png", size_hint=(coin_
            width, coin_height), pos_hint={'x': x, 'y': y}, allow_
            stretch=True)
```

```
        curr_screen.ids['layout_lvl'+str(screen_num)].add_widget(coin,
        index=-1)
        curr_screen.coins_ids['coin'+str(k)] = coin
class MainScreen(kivy.uix.screenmanager.Screen):
    pass

class Level1(kivy.uix.screenmanager.Screen):
    character_killed = False
    num_coins = 5
    num_coins_collected = 0
    coins_ids = {}

class Level2(kivy.uix.screenmanager.Screen):
    character_killed = False
    num_coins = 8
    num_coins_collected = 0
    coins_ids = {}

app = TestApp()
app.run()
```

Referencing Screens in Python Using Their Index

The reference to the current screen class is returned in the curr_screen variable using the next line. The keyword self refers to the TestApp class. root returns the root widget, which is the ScreenManager. In order to access the screens inside the manager, the screens list inside the ScreenManager is returned.

```
curr_screen = self.root.screens[screen_num]
```

When printed, the content of this list is shown in the next line where the screen's names are displayed.

```
[<Screen name='main'>, <Screen name='level1'>, <Screen name='level2'>]
```

The screen called level1 is the second element and thus has an index of 1. The level2 screen has an index of 2 and the main screen has an index of 0. After returning the class reference, we can access any variable within it. Figure 6-2 shows the level2 screen after running the application.

Figure 6-2. *Screen of the second level*

Screen on_enter Event

We replaced the build() method with the callback of the on_pre_enter event of the screen class. According to the Kivy application lifecycle, the next method to be called in our previous game is the on_start() method. Again, this method exists once for the entire application. We need to use a method that can differentiate between the different screens.

The Screen class has an event called on_enter that is fired exactly before the screen is displayed. In order to bind that event to the screens, the modified KV file is shown in Listing 6-8. Similar to the on_pre_enter event, there is a single callback function named screen_on_enter() associated with all screens. It accepts a number referring to the index of the screen firing the event.

Listing 6-8. Binding the on_enter Event to the Screens Inside the KV File

```
ScreenManager:
    MainScreen:
    Level1:
    Level2:
```

```
<MainScreen>:
    name: "main"
    BoxLayout:
        Button:
            text: "Go to Level 1"
            on_press: app.root.current="level1"
        Button:
            text: "Go to Level 2"
            on_press: app.root.current="level2"

<Level1>:
    name: "level1"
    on_pre_enter: app.screen_on_pre_enter(1)
    on_enter: app.screen_on_enter(1)
    FloatLayout:
        id: layout_lvl1
        on_touch_down: app.touch_down_handler(1, args)
        canvas.before:
            Rectangle:
                size: self.size
                pos: self.pos
                source: "bg_lvl1.jpg"
        NumCollectedCoins:
            id: num_coins_collected_lvl1
        Monster:
            id: monster_image_lvl1
        Character:
            id: character_image_lvl1

<Level2>:
    name: "level2"
    on_pre_enter: app.screen_on_pre_enter(2)
    on_enter: app.screen_on_enter(2)
    FloatLayout:
        id: layout_lvl2
        on_touch_down: app.touch_down_handler(2, args)
```

```
        canvas.before:
            Rectangle:
                size: self.size
                pos: self.pos
                source: "bg_lvl2.jpg"
            NumCollectedCoins:
                id: num_coins_collected_lvl2
            Monster:
                id: monster_image_lvl2
            Character:
                id: character_image_lvl2

<NumCollectedCoins@Label>:
    size_hint: (0.1, 0.02)
    pos_hint: {'x': 0.0, 'y': 0.97}
    text: "Coins 0"
    font_size: 20

<Monster@Image>:
    size_hint: (0.15, 0.15)
    pos_hint: {'x': 0.8, 'y': 0.8}
    source: "10.png"
    im_num: 10
    allow_stretch: True
    on_im_num: app.change_monst_im(self)
    on_pos_hint: app.monst_pos_hint(self)

<Character@Image>:
    size_hint: (0.15, 0.15)
    pos_hint: {'x': 0.2, 'y': 0.6}
    source: "0.png"
    im_num: 0
    allow_stretch: True
    on_im_num: app.change_char_im(self)
    on_pos_hint: app.char_pos_hint(self)
```

Listing 6-9 shows the Python file after implementing the screen_on_enter() callback function. After referencing the current screen and the Monster widget using its ID (by appending the screen number to the monster_image_lvl string), everything is similar to the on_start() method of the previous game.

Listing 6-9. Implementing the screen_on_enter() Callback Function Inside the Python File

```python
import kivy.app
import kivy.uix.screenmanager
import random

class TestApp(kivy.app.App):

    def screen_on_pre_enter(self, screen_num):
        coin_width = 0.05
        coin_height = 0.05

        curr_screen = self.root.screens[screen_num]

        section_width = 1.0/curr_screen.num_coins
        for k in range(curr_screen.num_coins):
            x = random.uniform(section_width*k, section_width*(k+1)-coin_
            width)
            y = random.uniform(0, 1-coin_height)
            coin = kivy.uix.image.Image(source="coin.png", size_hint=(coin_
            width, coin_height), pos_hint={'x': x, 'y': y}, allow_
            stretch=True)
            self.root.screens[screen_num].ids['layout_lvl'+str(screen_
            num)].add_widget(coin, index=-1)
            self.root.screens[screen_num].coins_ids['coin'+str(k)] = coin

    def screen_on_enter(self, screen_num):
        curr_screen = self.root.screens[screen_num]
        monster_image = curr_screen.ids['monster_image_lvl'+str(screen_num)]
        new_pos = (random.uniform(0.0, 1 - monster_image.size_hint[0]),
        random.uniform(0.0, 1 - monster_image.size_hint[1]))
        self.start_monst_animation(monster_image=monster_image, new_
        pos=new_pos, anim_duration=random.uniform(1.5, 3.5))
```

227

```python
class MainScreen(kivy.uix.screenmanager.Screen):
    pass

class Level1(kivy.uix.screenmanager.Screen):
    character_killed = False
    num_coins = 5
    num_coins_collected = 0
    coins_ids = {}

class Level2(kivy.uix.screenmanager.Screen):
    character_killed = False
    num_coins = 8
    num_coins_collected = 0
    coins_ids = {}

app = TestApp()
app.run()
```

Monster Animation

The screen_on_enter() function calls the start_monst_animation() function in order to start the animation of the monster. This function is implemented according to the modified Python file in Listing 6-10. In the previous game, the animation starts by referring to the monster Image widget using its predefined ID in the current multi-level game. Thus, this function is modified in order to accept the Monster widget for running the animation.

Listing 6-10. Starting the Monster Animation Inside the screen_on_enter() callback Function

```python
import kivy.app
import kivy.uix.screenmanager
import random

class TestApp(kivy.app.App):

    def screen_on_pre_enter(self, screen_num):
        coin_width = 0.05
        coin_height = 0.05
```

```
    curr_screen = self.root.screens[screen_num]

    section_width = 1.0/curr_screen.num_coins
    for k in range(curr_screen.num_coins):
        x = random.uniform(section_width*k, section_width*(k+1)-coin_
        width)
        y = random.uniform(0, 1-coin_height)
        coin = kivy.uix.image.Image(source="coin.png", size_hint=(coin_
        width, coin_height), pos_hint={'x': x, 'y': y}, allow_
        stretch=True)
        curr_screen.ids['layout_lvl'+str(screen_num)].add_widget(coin,
        index=-1)
        curr_screen.coins_ids['coin'+str(k)] = coin

def screen_on_enter(self, screen_num):
    curr_screen = self.root.screens[screen_num]
    monster_image = curr_screen.ids['monster_image_lvl'+str(screen_num)]
    new_pos = (random.uniform(0.0, 1 - monster_image.size_hint[0]),
    random.uniform(0.0, 1 - monster_image.size_hint[1]))
    self.start_monst_animation(monster_image=monster_image, new_
    pos=new_pos, anim_duration=random.uniform(1.5, 3.5))

def start_monst_animation(self, monster_image, new_pos, anim_duration):
    monst_anim = kivy.animation.Animation(pos_hint={'x': new_pos[0],
    'y': new_pos[1]}, im_num=17,duration=anim_duration)
    monst_anim.bind(on_complete=self.monst_animation_completed)
    monst_anim.start(monster_image)

def monst_animation_completed(self, *args):
    monster_image = args[1]
    monster_image.im_num = 10

    new_pos = (random.uniform(0.0, 1 - monster_image.size_hint[0]), random.
    uniform(0.0, 1 - monster_image.size_hint[1]))
    self.start_monst_animation(monster_image=monster_image, new_pos=new_
    pos, anim_duration=random.uniform(1.5, 3.5))
```

```
class MainScreen(kivy.uix.screenmanager.Screen):
    pass

class Level1(kivy.uix.screenmanager.Screen):
    character_killed = False
    num_coins = 5
    num_coins_collected = 0
    coins_ids = {}

class Level2(kivy.uix.screenmanager.Screen):
    character_killed = False
    num_coins = 8
    num_coins_collected = 0
    coins_ids = {}

app = TestApp()
app.run()
```

The Monster animation fires the on_complete event after being completed and has a callback function named monst_animaiton_completed() that's implemented in Listing 6-10.

Notice how the Monster widget is referenced. This event passes arguments inside the args variable. When it's printed, it looks as shown in this output:

(<kivy.animation.Animation object at 0x7f9eeb0f2a70>, <WeakProxy to <kivy. factory.Monster object at 0x7f9eeb15f8d0>>)

It is a tuple in which the first element at index 0 refers to the animation that fired the event and the second element at index 1 is the widget associated with the animation. Thus, we can reference the widget directly by indexing args with index 1.

Handling Monster on_pos_hint and on_im_num Events

According to the KV file in Listing 6-8, the Monster widget fires the on_pos_hint and the on_im_num events, which are handled using the monst_pos_hint() and change_monst_im() callback functions. Their implementation is shown in the new KV file in Listing 6-11.

Both of these functions accept the widget that fired the event. Inside the function, we need not only the Monster widget but also the Character widget. How do we reference that widget?

Listing 6-11. Handling the on_pos_hint and on_im_num Events of the Monster Widget

```
import kivy.app
import kivy.uix.screenmanager
import random
import kivy.clock
import functools

class TestApp(kivy.app.App):

    def screen_on_pre_enter(self, screen_num):
        coin_width = 0.05
        coin_height = 0.05

        curr_screen = self.root.screens[screen_num]

        section_width = 1.0/curr_screen.num_coins
        for k in range(curr_screen.num_coins):
            x = random.uniform(section_width*k, section_width*(k+1)-coin_
            width)
            y = random.uniform(0, 1-coin_height)
            coin = kivy.uix.image.Image(source="coin.png", size_hint=(coin_
            width, coin_height), pos_hint={'x': x, 'y': y}, allow_
            stretch=True)
            curr_screen.ids['layout_lvl'+str(screen_num)].add_widget(coin,
            index=-1)
            curr_screen.coins_ids['coin'+str(k)] = coin

    def screen_on_enter(self, screen_num):
        curr_screen = self.root.screens[screen_num]
        monster_image = curr_screen.ids['monster_image_lvl'+str(screen_num)]
        new_pos = (random.uniform(0.0, 1 - monster_image.size_hint[0]),
        random.uniform(0.0, 1 - monster_image.size_hint[1]))
        self.start_monst_animation(monster_image=monster_image, new_
        pos=new_pos, anim_duration=random.uniform(1.5, 3.5))
```

231

```python
    def start_monst_animation(self, monster_image, new_pos, anim_duration):
        monst_anim = kivy.animation.Animation(pos_hint={'x': new_pos[0],
        'y': new_pos[1]}, im_num=17,duration=anim_duration)
        monst_anim.bind(on_complete=self.monst_animation_completed)
        monst_anim.start(monster_image)

    def monst_animation_completed(self, *args):
        monster_image = args[1]
        monster_image.im_num = 10

        new_pos = (random.uniform(0.0, 1 - monster_image.size_hint[0]),
        random.uniform(0.0, 1 - monster_image.size_hint[1]))
        self.start_monst_animation(monster_image=monster_image, new_pos=
        new_pos,anim_duration=random.uniform(1.5, 3.5))

    def monst_pos_hint(self, monster_image):
        screen_num = int(monster_image.parent.parent.name[5:])
        curr_screen = self.root.screens[screen_num]
        character_image = curr_screen.ids['character_image_lvl'+str(screen_
        num)]

        character_center = character_image.center
        monster_center = monster_image.center

        gab_x = character_image.width / 2
        gab_y = character_image.height / 2
        if character_image.collide_widget(monster_image) and abs(character_
        center[0] - monster_center[0]) <= gab_x and abs(character_
        center[1] - monster_center[1]) <= gab_y and curr_screen.character_
        killed == False:
            curr_screen.character_killed = True

            kivy.animation.Animation.cancel_all(character_image)
            kivy.animation.Animation.cancel_all(monster_image)

            character_image.im_num = 91
            char_anim = kivy.animation.Animation(im_num=95)
            char_anim.start(character_image)
```

```
        kivy.clock.Clock.schedule_once(functools.partial(self.back_to_
        main_screen, curr_screen.parent), 3)
    def change_monst_im(self, monster_image):
        screen_num = int(monster_image.parent.parent.name[5:])
        monster_image.source = str(int(monster_image.im_num)) + ".png"

    def back_to_main_screen(self, screenManager, *args):
        screenManager.current = "main"

class MainScreen(kivy.uix.screenmanager.Screen):
    pass

class Level1(kivy.uix.screenmanager.Screen):
    character_killed = False
    num_coins = 5
    num_coins_collected = 0
    coins_ids = {}

class Level2(kivy.uix.screenmanager.Screen):
    character_killed = False
    num_coins = 8
    num_coins_collected = 0
    coins_ids = {}

app = TestApp()
app.run()
```

Referencing Screen FloatLayout Using Its Children

We can reference the parent FloatLayout of the Monster widget using its parent
property. From the Floatlayout, we can use the parent property again for referencing
its parent screen. After referencing the screen, we can access its name using the name
property. In the KV file, the names of the two levels screens have a number at the end
that refers to the screen index. The line that returns the screen index is shown below. The
index starts at 5.

In the current game with just two levels, the indices of the screens are 1 and 2. Thus, we can just use -1 rather than 5: to return the number at the end of the screen name. But this will work only if the index of the screen is a single number from 0 to 9. If the screen has index 10, it will not work.

In order to convert the index into a number, the int() function is used.

```
screen_num = int(monster_image.parent.parent.name[5:])
```

The returned index is saved in the screen_num variable. That index can be used to reference the Character widget based on its ID as done previously with the Monster widget. Just append it to the character_image_lvl string.

Returning to Main Screen

After the character is killed, the application routes to the main screen. This is done by scheduling to change the current property of the ScreenManager to change to main after three seconds using the kivy.clock.Clock.schedule_interval() callback function. A function named back_to_main_screen() is called and it accepts the ScreenManager. In order to pass arguments to the schedule_interval() function, we use the functools.partial() function.

At this point, we finished all required work for the Monster widget. Let's move on to the Character widget.

Handling Character Motion Using on_touch_down Event

According to the KV file in Listing 6-8, the FloatLayout has an event called on_touch_down that is handled using the touch_down_handler() callback function, as shown in the next Python file. This function accepts the screen index in the screen_num argument. It uses that index to reference the character_killed property defined in the screen class.

In order to start the animation, the callback function calls the start_char_animation() function that starts the animation of the Character widget. It is implemented in Listing 6-12. This function accepts the screen_num argument that represents the screen index. Its implementation is identical to the one discussed in the previous game, except for referencing the Character widget using the screen_num argument.

Listing 6-12. Implementing the start_char_animation() Callback Function to
Start the Animation of the Character Widget

```
import kivy.app
import kivy.uix.screenmanager
import random
import kivy.clock
import functools

class TestApp(kivy.app.App):

    def screen_on_pre_enter(self, screen_num):
        coin_width = 0.05
        coin_height = 0.05

        curr_screen = self.root.screens[screen_num]

        section_width = 1.0/curr_screen.num_coins
        for k in range(curr_screen.num_coins):
            x = random.uniform(section_width*k, section_width*(k+1)-coin_
            width)
            y = random.uniform(0, 1-coin_height)
            coin = kivy.uix.image.Image(source="coin.png", size_hint=(coin_
            width, coin_height), pos_hint={'x': x, 'y': y}, allow_
            stretch=True)
            curr_screen.ids['layout_lvl'+str(screen_num)].add_widget(coin,
            index=-1)
            curr_screen.coins_ids['coin'+str(k)] = coin

    def screen_on_enter(self, screen_num):
        curr_screen = self.root.screens[screen_num]
        monster_image = curr_screen.ids['monster_image_lvl'+str(screen_num)]
        new_pos = (random.uniform(0.0, 1 - monster_image.size_hint[0]),
        random.uniform(0.0, 1 - monster_image.size_hint[1]))
        self.start_monst_animation(monster_image=monster_image, new_
        pos=new_pos, anim_duration=random.uniform(1.5, 3.5))
```

```python
def start_monst_animation(self, monster_image, new_pos, anim_duration):
    monst_anim = kivy.animation.Animation(pos_hint={'x': new_pos[0],
    'y': new_pos[1]}, im_num=17,duration=anim_duration)
    monst_anim.bind(on_complete=self.monst_animation_completed)
    monst_anim.start(monster_image)

def monst_animation_completed(self, *args):
    monster_image = args[1]
    monster_image.im_num = 10

    new_pos = (random.uniform(0.0, 1 - monster_image.size_hint[0]),
    random.uniform(0.0, 1 - monster_image.size_hint[1]))
    self.start_monst_animation(monster_image=monster_image, new_pos=
    new_pos,anim_duration=random.uniform(1.5, 3.5))

def monst_pos_hint(self, monster_image):
    screen_num = int(monster_image.parent.parent.name[5:])
    curr_screen = self.root.screens[screen_num]
    character_image = curr_screen.ids['character_image_lvl'+str(screen_
    num)]

    character_center = character_image.center
    monster_center = monster_image.center

    gab_x = character_image.width / 2
    gab_y = character_image.height / 2
    if character_image.collide_widget(monster_image) and abs(character_
    center[0] - monster_center[0]) <= gab_x and abs(character_
    center[1] - monster_center[1]) <= gab_y and curr_screen.character_
    killed == False:
        curr_screen.character_killed = True

        kivy.animation.Animation.cancel_all(character_image)
        kivy.animation.Animation.cancel_all(monster_image)

        character_image.im_num = 91
        char_anim = kivy.animation.Animation(im_num=95)
        char_anim.start(character_image)
```

```
        kivy.clock.Clock.schedule_once(functools.partial(self.back_to_
            main_screen, curr_screen.parent), 3)

    def change_monst_im(self, monster_image):
        monster_image.source = str(int(monster_image.im_num)) + ".png"

    def touch_down_handler(self, screen_num, args):
        curr_screen = self.root.screens[screen_num]
        if curr_screen.character_killed == False:
            self.start_char_animation(screen_num, args[1].spos)

    def start_char_animation(self, screen_num, touch_pos):
        curr_screen = self.root.screens[screen_num]
        character_image = curr_screen.ids['character_image_lvl'+str(screen_
            num)]
        character_image.im_num = 0
        char_anim = kivy.animation.Animation(pos_hint={'x': touch_pos[0] -
            character_image.size_hint[0] / 2,'y': touch_pos[1] - character_
            image.size_hint[1] / 2}, im_num=7)
        char_anim.bind(on_complete=self.char_animation_completed)
        char_anim.start(character_image)

    def char_animation_completed(self, *args):
        character_image = args[1]
        character_image.im_num = 0

    def back_to_main_screen(self, screenManager, *args):
        screenManager.current = "main"

class MainScreen(kivy.uix.screenmanager.Screen):
    pass

class Level1(kivy.uix.screenmanager.Screen):
    character_killed = False
    num_coins = 5
    num_coins_collected = 0
    coins_ids = {}
```

```
class Level2(kivy.uix.screenmanager.Screen):
    character_killed = False
    num_coins = 8
    num_coins_collected = 0
    coins_ids = {}

app = TestApp()
app.run()
```

Because the on_complete event is bound to the animation, its callback function char_animation_completed() is implemented in the Python file in Listing 6-12. It references the Character widget using the args argument, as discussed before.

Handling Character on_pos_hint and on_im_num Events

Similar to the Monster widget, the Character widget defined in the KV file fires two events—on_pos_hint and on_im_num. Their callback functions are implemented as shown in the Python file in Listing 6-13, which is similar to what discussed previously.

When all the coins are collected, the application returns to the main screen, as done previously when the character is killed.

Listing 6-13. Handling the on_pos_hint and on_im_num Events of the Character Widget

```
import kivy.app
import kivy.uix.screenmanager
import random
import kivy.clock
import functools

class TestApp(kivy.app.App):

    def screen_on_pre_enter(self, screen_num):
        coin_width = 0.05
        coin_height = 0.05
```

```python
        curr_screen = self.root.screens[screen_num]

        section_width = 1.0/curr_screen.num_coins
        for k in range(curr_screen.num_coins):
            x = random.uniform(section_width*k, section_width*(k+1)-coin_
            width)
            y = random.uniform(0, 1-coin_height)
            coin = kivy.uix.image.Image(source="coin.png", size_hint=(coin_
            width, coin_height), pos_hint={'x': x, 'y': y}, allow_
            stretch=True)
            curr_screen.ids['layout_lvl'+str(screen_num)].add_widget(coin,
            index=-1)
            curr_screen.coins_ids['coin'+str(k)] = coin

    def screen_on_enter(self, screen_num):
        curr_screen = self.root.screens[screen_num]
        monster_image = curr_screen.ids['monster_image_lvl'+str(screen_num)]
        new_pos = (random.uniform(0.0, 1 - monster_image.size_hint[0]),
        random.uniform(0.0, 1 - monster_image.size_hint[1]))
        self.start_monst_animation(monster_image=monster_image, new_
        pos=new_pos, anim_duration=random.uniform(1.5, 3.5))

    def start_monst_animation(self, monster_image, new_pos, anim_duration):
        monst_anim = kivy.animation.Animation(pos_hint={'x': new_pos[0],
        'y': new_pos[1]}, im_num=17,duration=anim_duration)
        monst_anim.bind(on_complete=self.monst_animation_completed)
        monst_anim.start(monster_image)

    def monst_animation_completed(self, *args):
        monster_image = args[1]
        monster_image.im_num = 10

        new_pos = (random.uniform(0.0, 1 - monster_image.size_hint[0]),
        random.uniform(0.0, 1 - monster_image.size_hint[1]))
        self.start_monst_animation(monster_image=monster_image, new_pos=
        new_pos,anim_duration=random.uniform(1.5, 3.5))
```

```python
def monst_pos_hint(self, monster_image):
    screen_num = int(monster_image.parent.parent.name[5:])
    curr_screen = self.root.screens[screen_num]
    character_image = curr_screen.ids['character_image_lvl'+str(screen_
num)]

    character_center = character_image.center
    monster_center = monster_image.center

    gab_x = character_image.width / 2
    gab_y = character_image.height / 2
    if character_image.collide_widget(monster_image) and abs(character_
center[0] - monster_center[0]) <= gab_x and abs(character_
center[1] - monster_center[1]) <= gab_y and curr_screen.character_
killed == False:
        curr_screen.character_killed = True

        kivy.animation.Animation.cancel_all(character_image)
        kivy.animation.Animation.cancel_all(monster_image)

        character_image.im_num = 91
        char_anim = kivy.animation.Animation(im_num=95)
        char_anim.start(character_image)
        kivy.clock.Clock.schedule_once(functools.partial(self.back_to_
main_screen, curr_screen.parent), 3)

def change_monst_im(self, monster_image):
    monster_image.source = str(int(monster_image.im_num)) + ".png"

def touch_down_handler(self, screen_num, args):
    curr_screen = self.root.screens[screen_num]
    if curr_screen.character_killed == False:
        self.start_char_animation(screen_num, args[1].spos)

def start_char_animation(self, screen_num, touch_pos):
    curr_screen = self.root.screens[screen_num]
    character_image = curr_screen.ids['character_image_lvl'+str(screen_
num)]
    character_image.im_num = 0
```

```
    char_anim = kivy.animation.Animation(pos_hint={'x': touch_pos[0] -
    character_image.size_hint[0] / 2,'y': touch_pos[1] - character_
    image.size_hint[1] / 2}, im_num=7)
    char_anim.bind(on_complete=self.char_animation_completed)
    char_anim.start(character_image)

def char_animation_completed(self, *args):
    character_image = args[1]
    character_image.im_num = 0

def char_pos_hint(self, character_image):
    screen_num = int(character_image.parent.parent.name[5:])
    character_center = character_image.center

    gab_x = character_image.width / 3
    gab_y = character_image.height / 3
    coins_to_delete = []
    curr_screen = self.root.screens[screen_num]

    for coin_key, curr_coin in curr_screen.coins_ids.items():
        curr_coin_center = curr_coin.center
        if character_image.collide_widget(curr_coin) and abs(character_
        center[0] - curr_coin_center[0]) <= gab_x and abs(character_
        center[1] - curr_coin_center[1]) <= gab_y:
            coins_to_delete.append(coin_key)
            curr_screen.ids['layout_lvl'+str(screen_num)].remove_
            widget(curr_coin)
            curr_screen.num_coins_collected = curr_screen.num_coins_
            collected + 1
            curr_screen.ids['num_coins_collected_lvl'+str(screen_num)].
            text = "Coins "+str(curr_screen.num_coins_collected)
            if curr_screen.num_coins_collected == curr_screen.num_
            coins:
                kivy.animation.Animation.cancel_all(character_image)
                kivy.clock.Clock.schedule_once(functools.partial(self.
                back_to_main_screen, curr_screen.parent), 3)
                kivy.animation.Animation.cancel_all(curr_screen.
                ids['monster_image_lvl'+str(screen_num)])
```

241

```
                    curr_screen.ids['layout_lvl'+str(screen_num)].add_
                    widget(kivy.uix.label.Label(pos_hint={'x': 0.1, 'y':
                    0.1}, size_hint=(0.8, 0.8), font_size=90, text="Level
                    Completed"))

        if len(coins_to_delete) > 0:
            for coin_key in coins_to_delete:
                del curr_screen.coins_ids[coin_key]

    def change_char_im(self, character_image):
        character_image.source = str(int(character_image.im_num)) + ".png"

    def back_to_main_screen(self, screenManager, *args):
        screenManager.current = "main"

class MainScreen(kivy.uix.screenmanager.Screen):
    pass

class Level1(kivy.uix.screenmanager.Screen):
    character_killed = False
    num_coins = 5
    num_coins_collected = 0
    coins_ids = {}

class Level2(kivy.uix.screenmanager.Screen):
    character_killed = False
    num_coins = 8
    num_coins_collected = 0
    coins_ids = {}

app = TestApp()
app.run()
```

As of the previous step, a two-level game has been successfully created that uses almost the same code used in the previous single-level game. But there are some issues that will happen when playing it. Did you catch any of the issues? Don't worry; we will discuss them in the next section.

Issues with the Game

There are six issues that exist in the previous game. Most of them are due to the improper initialization of the variables defined in the Screen classes (i.e., Level1 and Level2).

Issue 1: Character Dies Immediately After the Level Starts

The first issue occurs when the character is killed when playing a level and then we try to start that level again. The character can easily be killed as soon as the level starts again, depending on the motion of the monster. Let's discuss this issue.

When the position of the monster and the character satisfies the condition inside the monst_pos_hint() callback function, this means a collision occurred between the character and the monster and thus the character is killed. The application will be redirected automatically to the main screen after three seconds. When players start the same level again, the character and monster exist in the same location where the character was killed. This might lead to killing the character immediately after the new level starts, as the player will not be able to move the character fast enough.

In the case shown in Figure 6-3, the character and the monster are moving in the directions marked by the arrows. Because they are moving in different directions, the new position of the monster might be away from the current character position. As a result, the character will not be killed. Regardless, we have to solve this problem.

Figure 6-3. *The character and monster widgets are moving away from each other*

Solution

After the level ends, we have to reset the positions of the character and the monster. In fact, we have to reset everything calculated from the previous play time, such as the number of coins collected.

It is best to reset the positions after the level ends, not before starting the level again, to make sure there is no chance for collision. The Screen class provides an event called on_pre_leave that is fired before leaving the screen. Within its callback function, we can change the positions of the character and the monster to make sure they are away from each other when the next level starts.

The modified KV file for binding the on_pre_leave to both screens is shown in Listing 6-14. It is essential to bind this event to each level screen. The callback function named screen_on_pre_leave() is attached to this event. It accepts an argument that defines the screen index in order to access the screen and its widgets within the Python file.

Listing 6-14. Binding the on_pre_leave to the Screens

```
ScreenManager:
    MainScreen:
    Level1:
    Level2:

<MainScreen>:
    name: "main"
    BoxLayout:
        Button:
            text: "Go to Level 1"
            on_press: app.root.current="level1"
        Button:
            text: "Go to Level 2"
            on_press: app.root.current="level2"

<Level1>:
    name: "level1"
    on_pre_enter: app.screen_on_pre_enter(1)
    on_pre_leave: app.screen_on_pre_leave(1)
    on_enter: app.screen_on_enter(1)
    FloatLayout:
        id: layout_lvl1
        on_touch_down: app.touch_down_handler(1, args)
        canvas.before:
```

```
        Rectangle:
            size: self.size
            pos: self.pos
            source: "bg_lvl1.jpg"
        NumCollectedCoins:
            id: num_coins_collected_lvl1
        Monster:
            id: monster_image_lvl1
        Character:
            id: character_image_lvl1

<Level2>:
    name: "level2"
    on_pre_enter: app.screen_on_pre_enter(2)
    on_pre_leave: app.screen_on_pre_leave(2)
    on_enter: app.screen_on_enter(2)
    FloatLayout:
        id: layout_lvl2
        on_touch_down: app.touch_down_handler(2, args)
        canvas.before:
            Rectangle:
                size: self.size
                pos: self.pos
                source: "bg_lvl2.jpg"
        NumCollectedCoins:
            id: num_coins_collected_lvl2
        Monster:
            id: monster_image_lvl2
        Character:
            id: character_image_lvl2

<NumCollectedCoins@Label>:
    size_hint: (0.1, 0.02)
    pos_hint: {'x': 0.0, 'y': 0.97}
    text: "Coins 0"
    font_size: 20
```

```
<Monster@Image>:
    size_hint: (0.15, 0.15)
    pos_hint: {'x': 0.8, 'y': 0.8}
    source: "10.png"
    im_num: 10
    allow_stretch: True
    on_im_num: app.change_monst_im(self)
    on_pos_hint: app.monst_pos_hint(self)

<Character@Image>:
    size_hint: (0.15, 0.15)
    pos_hint: {'x': 0.2, 'y': 0.6}
    source: "0.png"
    im_num: 0
    allow_stretch: True
    on_im_num: app.change_char_im(self)
    on_pos_hint: app.char_pos_hint(self)
```

The implementation of this function is shown in Listing 6-15. It sets the positions of the monster and the character to the same positions used inside the KV file.

Listing 6-15. Implementing the screen_on_pre_leave() Callback Function

```
def screen_on_pre_leave(self, screen_num):
    curr_screen = self.root.screens[screen_num]

    curr_screen.ids['monster_image_lvl' + str(screen_num)].pos_hint = {'x':
    0.8, 'y': 0.8}
    curr_screen.ids['character_image_lvl' + str(screen_num)].pos_hint =
    {'x': 0.2, 'y': 0.6}
```

Issue 2: Character Does Not Move After Restarting the Same Level

After the character is killed, the monst_pos_hint() callback functions sets the character_killed class variable to True. This flag remains true even after restarting the same level. According to the touch_down_handler() callback function in the Python file, it only starts the character animation if this flag is False.

Solution

In order to solve this issue, the character_killed flag must be reset to False. This allows the touch_down_handler() function to animate the character to move to the touched position.

Because this value must be false before the screen starts, we can reset it inside the screen_on_pre_enter() callback function of the on_pre_enter event. The modified function is shown in Listing 6-16. This allows the character to move when restarting the same level in which the character was killed.

Listing 6-16. Allowing the Character to Move When Replaying a Level It Was Killed in Previously

```python
def screen_on_pre_enter(self, screen_num):
    curr_screen = self.root.screens[screen_num]
    curr_screen.character_killed = False

    coin_width = 0.05
    coin_height = 0.05

    section_width = 1.0 / curr_screen.num_coins
    for k in range(curr_screen.num_coins):
        x = random.uniform(section_width * k, section_width * (k + 1) -
        coin_width)
        y = random.uniform(0, 1 - coin_height)
        coin = kivy.uix.image.Image(source="coin.png", size_hint=(coin_
        width, coin_height), pos_hint={'x': x, 'y': y},
                                    allow_stretch=True)
        curr_screen.ids['layout_lvl' + str(screen_num)].add_widget(coin,
        index=-1)
        curr_screen.coins_ids['coin' + str(k)] = coin
```

Issue 3: Character Image Starts with a Dead Image

After the character is killed, there is an animation at the end of the monst_pos_hint() callback function that changes its im_num from 91 to 95 before going to the main screen. After the animation, the im_num property will be set to 95.

If the level is restarted, the value of im_num will still be 95, which is the dead character image. We can reset the monster image the same way.

Solution

The solution is to change the value of the character's im_num property to 0, which refers to an alive character. The monster image is also reset to 0. This can be changed inside the screen_on_pre_enter() callback function, which is modified in Listing 6-17.

Listing 6-17. Forcing the im_num property of the Character Widget to 0 Before Starting a Level

```
def screen_on_pre_enter(self, screen_num):
    curr_screen = self.root.screens[screen_num]
    curr_screen.character_killed = False
    curr_screen.ids['character_image_lvl' + str(screen_num)].im_num = 0
    curr_screen.ids['monster_image_lvl' + str(screen_num)].im_num = 10

    coin_width = 0.05
    coin_height = 0.05

    section_width = 1.0 / curr_screen.num_coins
    for k in range(curr_screen.num_coins):
        x = random.uniform(section_width * k, section_width * (k + 1) -
        coin_width)
        y = random.uniform(0, 1 - coin_height)
        coin = kivy.uix.image.Image(source="coin.png", size_hint=(coin_
        width, coin_height), pos_hint={'x': x, 'y': y},
                                    allow_stretch=True)
        curr_screen.ids['layout_lvl' + str(screen_num)].add_widget(coin,
        index=-1)
        curr_screen.coins_ids['coin' + str(k)] = coin
```

Issue 4: Uncollected Coins Remain in the Next Trial of the Same Level

When the character is killed, there will be some remaining coins that were not collected during that level. Unfortunately, these coins will remain as uncollected when players try to replay the same level. Let's discuss why this occurs.

Assume that we selected the second level from the main screen, which has eight coins according to num_coins defined in the class. Before the screen starts, the coins are created in the callback function of the on_pre_enter event, which is screen_on_pre_enter(). Within that function, a new Image widget for each coins is created and added as a child widget to the FloatLayout. In order to reference the added widgets, the coins are also added inside the coins_ids dictionary defined in the Screen class. The content of the layout and the dictionary before and after executing that callback function are printed according to Figure 6-4. Note that the children within a widget are returned using the children property.

```
Layout Before on_pre_enter:  [<kivy.factory.Character object at 0x7f2b6a9ff048>, <kivy.fa
ctory.Monster object at 0x7f2b6aa8dee8>, <kivy.factory.NumCollectedCoins object at 0x7f2b
6aa8de80>]
Dictionary Before on pre enter:  {}
libpng warning: iCCP: known incorrect sRGB profile
Layout After on_pre_enter:  [<kivy.factory.Character object at 0x7f2b6a9ff048>, <kivy.fac
tory.Monster object at 0x7f2b6aa8dee8>, <kivy.uix.image.Image object at 0x7f2b615a65f8>,
<kivy.uix.image.Image object at 0x7f2b615a6798>, <kivy.uix.image.Image object at 0x7f2b61
5a6868>, <kivy.uix.image.Image object at 0x7f2b615a6938>, <kivy.uix.image.Image object at
 0x7f2b615a6a08>, <kivy.uix.image.Image object at 0x7f2b615a6ad8>, <kivy.uix.image.Image
object at 0x7f2b615a6ba8>, <kivy.uix.image.Image object at 0x7f2b615a6c78>, <kivy.factory
.NumCollectedCoins object at 0x7f2b6aa8de80>]
Dictionary After on_pre_enter:  {'coin0': <kivy.uix.image.Image object at 0x7f2b615a65f8>
, 'coin1': <kivy.uix.image.Image object at 0x7f2b615a6798>, 'coin2': <kivy.uix.image.Imag
e object at 0x7f2b615a6868>, 'coin3': <kivy.uix.image.Image object at 0x7f2b615a6938>, 'c
oin4': <kivy.uix.image.Image object at 0x7f2b615a6a08>, 'coin5': <kivy.uix.image.Image ob
ject at 0x7f2b615a6ad8>, 'coin6': <kivy.uix.image.Image object at 0x7f2b615a6ba8>, 'coin7
': <kivy.uix.image.Image object at 0x7f2b615a6c78>}
```

Figure 6-4. *Printing the coins dictionary and game Layout before and after calling the on_pre_enter() callback function*

Before on_pre_enter is fired, the dictionary is empty, as initialized in the class, and the layout has just three children, which are the ones added inside the KV file (Character, Monster, and NumCollectedCoins).

After the event is handled inside the screen_on_pre_enter() function, the eight coins Image widgets are created and added to the layout and dictionary. Thus, the layout will have a total of 11 children and the dictionary will have eight items.

Figure 6-5 displays the coins after being randomly distributed across the screen.

Figure 6-5. *Randomly distributing the coins across the screen*

Assume that the character is killed after collecting two of the eight coins, as shown in Figure 6-6. Note that the text currently displayed in the label at the top-left corner reflects that there is only two coins collected. As a result, there are six remaining coins in the screen.

Figure 6-6. *The character is killed after collecting two coins out of 8*

Players expect to have an empty dictionary and just three child widgets inside the layout before the on_pre_enter is fired, when they restart that level. But this does not happen according to Figure 6-7. Before handling the event, the layout has nine children rather than three and the dictionary has six children rather than zero.

```
Layout Before on_pre_enter: [<kivy.factory.Character object at 0x7f2b6a9ff048>, <kivy.fa
ctory.Monster object at 0x7f2b6aa8dee8>, <kivy.uix.image.Image object at 0x7f2b615a65f8>,
 <kivy.uix.image.Image object at 0x7f2b615a6798>, <kivy.uix.image.Image object at 0x7f2b6
15a6938>, <kivy.uix.image.Image object at 0x7f2b615a6ad8>, <kivy.uix.image.Image object a
t 0x7f2b615a6ba8>, <kivy.uix.image.Image object at 0x7f2b615a6c78>, <kivy.factory.NumColl
ectedCoins object at 0x7f2b6aa8de80>]
Dictionary Before on_pre_enter: {'coin0': <kivy.uix.image.Image object at 0x7f2b615a65f8
>, 'coin1': <kivy.uix.image.Image object at 0x7f2b615a6798>, 'coin3': <kivy.uix.image.Ima
ge object at 0x7f2b615a6938>, 'coin5': <kivy.uix.image.Image object at 0x7f2b615a6ad8>, '
coin6': <kivy.uix.image.Image object at 0x7f2b615a6ba8>, 'coin7': <kivy.uix.image.Image o
bject at 0x7f2b615a6c78>}
```

Figure 6-7. *The coins dictionary does not reset after replaying a level and stores the coin information from the previous trial*

After collecting two coins during the first time level 2 is played, the remaining six coins are not removed from the dictionary. Notice that the dictionary the keys—coin2 and coin4—and they are collected the previous time because their keys are missing from the dictionary.

The layout and the dictionary must be reset after the level ends.

Note that the game will work well without resetting the dictionary, but it is better to reset all class variables to their initial values. After the screen_on_pre_enter() callback function is executed, the content of the dictionary and the layout without being reset are shown in Figure 6-8. There are eight new Image widgets representing the new coins added to the layout and thus the total number of children within the layout is now 17. Of the 17 widgets, three widgets represent the Character, Monster, and NumCollectedCoins. The remaining six widgets are the coins Image widgets that were not collected the previous time level 2 was played.

The dictionary has eight items and that means the previous six coins were deleted from the dictionary. The previous six widgets were not deleted but overwritten by the new widgets, as the dictionary uses the same keys (coin0 to coin7) for the items. Notice that the items with keys coin2 and coin4, which are the coins collected previously, are added to the end of the dictionary.

```
Layout After on_pre_enter: [<kivy.factory.Character object at 0x7f2b6a9ff048>, <kivy.fac
tory.Monster object at 0x7f2b6aa8dee8>, <kivy.uix.image.Image object at 0x7f2b615a65f8>,
<kivy.uix.image.Image object at 0x7f2b615a6798>, <kivy.uix.image.Image object at 0x7f2b61
5a6938>, <kivy.uix.image.Image object at 0x7f2b615a6ad8>, <kivy.uix.image.Image object at
 0x7f2b615a6ba8>, <kivy.uix.image.Image object at 0x7f2b615a6c78>, <kivy.uix.image.Image
object at 0x7f2b61577798>, <kivy.uix.image.Image object at 0x7f2b615778d0>, <kivy.uix.ima
ge.Image object at 0x7f2b6157797a0>, <kivy.uix.image.Image object at 0x7f2b61577a70>, <kiv
y.uix.image.Image object at 0x7f2b61577b40>, <kivy.uix.image.Image object at 0x7f2b61577c
10>, <kivy.uix.image.Image object at 0x7f2b61577ce0>, <kivy.uix.image.Image object at 0x7
f2b61577db0>. <kivy.factory.NumCollectedCoins object at 0x7f2b6aa8de80>]
Dictionary After on pre enter: {'coin0': <kivy.uix.image.Image object at 0x7f2b61577798>
, 'coin1': <kivy.uix.image.Image object at 0x7f2b615778d0>, 'coin3': <kivy.uix.image.Imag
e object at 0x7f2b61577a70>, 'coin5': <kivy.uix.image.Image object at 0x7f2b61577c10>, 'c
oin6': <kivy.uix.image.Image object at 0x7f2b61577ce0>, 'coin7': <kivy.uix.image.Image ob
ject at 0x7f2b61577db0>, 'coin2': <kivy.uix.image.Image object at 0x7f2b6157797a0>, 'coin4
': <kivy.uix.image.Image object at 0x7f2b61577b40>}
```

Figure 6-8. *Coins not collected in the previous trial of a level are available when playing the level again*

When the screen is displayed, the result is shown in Figure 6-9. Notice that the previous six coins from the last time level 2 was played are displayed on the screen, as they exist in the FloatLayout. But these coins cannot be collected. That's because their position is retrieved from the coins_ids dictionary. Because the previous six coins were overwritten by the new eight widgets, we cannot access their positions. Thus, you are only permitted to select eight coins from the 14 coins displayed on the screen.

Figure 6-9. *Coins not collected from the previous trial of a level appear when replaying the level*

Solution

Because we need to make sure the coins_ids dictionary is empty and the layout has
no previous coins Image widgets, we can reset them inside the screen_on_pre_enter()
callback function.

The modified function is listed in Listing 6-18. In order to remove a widget from the
FloatLayout, we must reference the widget. Remember that all references to the coins
widgets from the previous trial are stored in the coins_ids dictionary. We can fetch each
dictionary item, return the coin widget, then pass it to the remove_widget() function,
which is called by the FloatLayout that is fetched using its ID and the screen_num
argument of the function. After the loop ends, the dictionary is reset using {}.

Listing 6-18. Resetting the Coins Dictionary and Layout Before Playing a Level

```
def screen_on_pre_enter(self, screen_num):
    curr_screen = self.root.screens[screen_num]
    curr_screen.character_killed = False
    curr_screen.ids['character_image_lvl' + str(screen_num)].im_num = 0
    curr_screen.ids['monster_image_lvl' + str(screen_num)].im_num = 10

    for key, coin in curr_screen.coins_ids.items():
        curr_screen.ids['layout_lvl' + str(screen_num)].remove_widget(coin)
    curr_screen.coins_ids = {}

    coin_width = 0.05
    coin_height = 0.05

    section_width = 1.0 / curr_screen.num_coins
    for k in range(curr_screen.num_coins):
        x = random.uniform(section_width * k, section_width * (k + 1) -
        coin_width)
        y = random.uniform(0, 1 - coin_height)
        coin = kivy.uix.image.Image(source="coin.png", size_hint=(coin_
        width, coin_height), pos_hint={'x': x, 'y': y},
                                    allow_stretch=True)
        curr_screen.ids['layout_lvl' + str(screen_num)].add_widget(coin,
        index=-1)
        curr_screen.coins_ids['coin' + str(k)] = coin
```

By doing that, no coins from a previous trial of a given level will appear in the current trial. We can also print the layout children and the dictionary items, as shown in Figure 6-10, to make sure everything works as expected. After starting the same level the second time, the layout has the three widgets defined in the KV file and no more. Also, the dictionary is empty.

```
Layout Before on pre enter: [<kivy.factory.Character object at 0x7ff36f7bb048>, <kivy.factory.Monst
er object at 0x7ff36f84aee8>, <kivy.factory.NumCollectedCoins object at 0x7ff36f84ae80>]
Dictionary Before on pre enter: {}
Layout After on pre enter: [<kivy.factory.Character object at 0x7ff36f7bb048>, <kivy.factory.Monste
r object at 0x7ff36f84aee8>, <kivy.uix.image.Image object at 0x7ff3663625f8>, <kivy.uix.image.Image
object at 0x7ff366362798>, <kivy.uix.image.Image object at 0x7ff366362868>, <kivy.uix.image.Image ob
ject at 0x7ff366362938>, <kivy.uix.image.Image object at 0x7ff366362a08>, <kivy.uix.image.Image obje
ct at 0x7ff366362ad8>, <kivy.uix.image.Image object at 0x7ff366362ba8>, <kivy.uix.image.Image object
at 0x7ff366362c78>, <kivy.factory.NumCollectedCoins object at 0x7ff36f84ae80>]
Dictionary After on pre enter: {'coin0': <kivy.uix.image.Image object at 0x7ff3663625f8>, 'coin1':
<kivy.uix.image.Image object at 0x7ff366362798>, 'coin2': <kivy.uix.image.Image object at 0x7ff36636
2868>, 'coin3': <kivy.uix.image.Image object at 0x7ff366362938>, 'coin4': <kivy.uix.image.Image obje
ct at 0x7ff366362a08>, 'coin5': <kivy.uix.image.Image object at 0x7ff366362ad8>, 'coin6': <kivy.uix.
image.Image object at 0x7ff366362ba8>, 'coin7': <kivy.uix.image.Image object at 0x7ff366362c78>}
Layout Before on pre enter: [<kivy.factory.Character object at 0x7ff36f7bb048>, <kivy.factory.Monst
er object at 0x7ff36f84aee8>, <kivy.factory.NumCollectedCoins object at 0x7ff36f84ae80>]
Dictionary Before on pre enter: {}
Layout After on pre enter: [<kivy.factory.Character object at 0x7ff36f7bb048>, <kivy.factory.Monste
r object at 0x7ff36f84aee8>, <kivy.uix.image.Image object at 0x7ff366332798>, <kivy.uix.image.Image
object at 0x7ff3663328d0>, <kivy.uix.image.Image object at 0x7ff3663329a0>, <kivy.uix.image.Image ob
ject at 0x7ff366332a70>, <kivy.uix.image.Image object at 0x7ff366332b40>, <kivy.uix.image.Image obje
ct at 0x7ff366332c10>, <kivy.uix.image.Image object at 0x7ff366332ce0>, <kivy.uix.image.Image object
at 0x7ff366332db0>, <kivy.factory.NumCollectedCoins object at 0x7ff36f84ae80>]
Dictionary After on pre enter: {'coin0': <kivy.uix.image.Image object at 0x7ff366332798>, 'coin1':
<kivy.uix.image.Image object at 0x7ff3663328d0>, 'coin2': <kivy.uix.image.Image object at 0x7ff36633
29a0>, 'coin3': <kivy.uix.image.Image object at 0x7ff366332a70>, 'coin4': <kivy.uix.image.Image obje
ct at 0x7ff366332b40>, 'coin5': <kivy.uix.image.Image object at 0x7ff366332c10>, 'coin6': <kivy.uix.
image.Image object at 0x7ff366332ce0>, 'coin7': <kivy.uix.image.Image object at 0x7ff366332db0>}
```

Figure 6-10. *Printing the layout children and the coins dictionary to make sure coins in a trial don't appear in the next trial*

Issue 5: The NumCollectedCoins Label Widget Doesn't Start with the "Coins 0" Text

Figure 6-11 shows the result when the character is killed after it collected two coins. The label displays "Coins 2" which confirms that two coins were collected.

Figure 6-11. *The character is killed after collecting two coins*

When the level is repeated, we expect to see the text `"Coins 0"` on the label, but `"Coins 2"` is displayed instead, as shown in Figure 6-12.

Figure 6-12. *The text in the label has not be reset*

Solution

In order to solve this issue, the displayed text on the label must be reset to "Coins 0" before the screen enters according to the modified screen_on_pre_enter() callback function in Listing 6-19.

Listing 6-19. Resetting the Label Text to "Coins 0" Before Entering a Screen

```python
def screen_on_pre_enter(self, screen_num):
    curr_screen = self.root.screens[screen_num]
    curr_screen.character_killed = False
    curr_screen.ids['character_image_lvl' + str(screen_num)].im_num = 0
    curr_screen.ids['monster_image_lvl' + str(screen_num)].im_num = 10
    curr_screen.ids['num_coins_collected_lvl' + str(screen_num)].text =
    "Coins 0"

    for key, coin in curr_screen.coins_ids.items():
        curr_screen.ids['layout_lvl' + str(screen_num)].remove_widget(coin)
    curr_screen.coins_ids = {}

    coin_width = 0.05
    coin_height = 0.05

    section_width = 1.0 / curr_screen.num_coins
    for k in range(curr_screen.num_coins):
        x = random.uniform(section_width * k, section_width * (k + 1) -
        coin_width)
        y = random.uniform(0, 1 - coin_height)
        coin = kivy.uix.image.Image(source="coin.png", size_hint=(coin_
        width, coin_height), pos_hint={'x': x, 'y': y},
                                    allow_stretch=True)
        curr_screen.ids['layout_lvl' + str(screen_num)].add_widget(coin,
        index=-1)
        curr_screen.coins_ids['coin' + str(k)] = coin
```

Issue 6: Number of Collected Coins Doesn't Start at 0 in the Next Trials

Resetting the text displayed on the Label widget does not solve the previous issue completely. Our target is to reset the number of collected coins to 0 when a level starts again. When a coin is collected, we expect the number to increase by 1 and the text to change from "Coins 0" to "Coins 1" but this does not happen. We did reset the text displayed once the level starts, but that does not solve the issue completely.

The number displayed on that label is taken from the num_coins_collected variable defined in the class. From Figure 6-12, the variable value is 2. Because its value is not reset, it will keep the same value when the level starts again. When a single coin is collected in the next trial, the number will be incremented by 1 to be 3,as displayed in Figure 6-13. Thus, we need to reset the variable to start at 0 in the next trials.

Figure 6-13. *The number of coins collected starts off from where the game stopped in the last trial*

Solution

Inside the screen_on_pre_enter() function in Listing 6-20, the value of the num_coins_ collected class variable is changed to 0.

Listing 6-20. Resetting the num_coins_collected to 0 Before Starting a Screen

```python
def screen_on_pre_enter(self, screen_num):
    curr_screen = self.root.screens[screen_num]
    curr_screen.character_killed = False
    curr_screen.num_coins_collected = 0
    curr_screen.ids['character_image_lvl' + str(screen_num)].im_num = 0
    curr_screen.ids['monster_image_lvl' + str(screen_num)].im_num = 10
    curr_screen.ids['num_coins_collected_lvl' + str(screen_num)].text =
    "Coins 0"

    for key, coin in curr_screen.coins_ids.items():
        curr_screen.ids['layout_lvl' + str(screen_num)].remove_widget(coin)
    curr_screen.coins_ids = {}

    coin_width = 0.05
    coin_height = 0.05

    curr_screen = self.root.screens[screen_num]

    section_width = 1.0 / curr_screen.num_coins
    for k in range(curr_screen.num_coins):
        x = random.uniform(section_width * k, section_width * (k + 1) -
        coin_width)
        y = random.uniform(0, 1 - coin_height)
        coin = kivy.uix.image.Image(source="coin.png", size_hint=(coin_
        width, coin_height), pos_hint={'x': x, 'y': y},
                                    allow_stretch=True)
        curr_screen.ids['layout_lvl' + str(screen_num)].add_widget(coin,
        index=-1)
        curr_screen.coins_ids['coin' + str(k)] = coin
```

Complete Game Implementation

After solving all issues in the application, the complete code of the game will be listed in this section. Listing 6-21 shows the KV file. The complete code for the Python file is shown in Listing 6-22.

KV File

Listing 6-21. Complete KV File of the Game

```
ScreenManager:
    MainScreen:
    Level1:
    Level2:

<MainScreen>:
    name: "main"
    BoxLayout:
        Button:
            text: "Go to Level 1"
            on_press: app.root.current="level1"
        Button:
            text: "Go to Level 2"
            on_press: app.root.current="level2"

<Level1>:
    name: "level1"
    on_pre_enter: app.screen_on_pre_enter(1)
    on_pre_leave: app.screen_on_pre_leave(1)
    on_enter: app.screen_on_enter(1)
    FloatLayout:
        id: layout_lvl1
        on_touch_down: app.touch_down_handler(1, args)
        canvas.before:
            Rectangle:
                size: self.size
                pos: self.pos
                source: "bg_lvl1.jpg"
        NumCollectedCoins:
            id: num_coins_collected_lvl1
        Monster:
            id: monster_image_lvl1
```

```
        Character:
            id: character_image_lvl1

<Level2>:
    name: "level2"
    on_pre_enter: app.screen_on_pre_enter(2)
    on_pre_leave: app.screen_on_pre_leave(2)
    on_enter: app.screen_on_enter(2)
    FloatLayout:
        id: layout_lvl2
        on_touch_down: app.touch_down_handler(2, args)
        canvas.before:
            Rectangle:
                size: self.size
                pos: self.pos
                source: "bg_lvl2.jpg"
        NumCollectedCoins:
            id: num_coins_collected_lvl2
        Monster:
            id: monster_image_lvl2
        Character:
            id: character_image_lvl2

<NumCollectedCoins@Label>:
    size_hint: (0.1, 0.02)
    pos_hint: {'x': 0.0, 'y': 0.97}
    text: "Coins 0"
    font_size: 20

<Monster@Image>:
    size_hint: (0.15, 0.15)
    pos_hint: {'x': 0.8, 'y': 0.8}
    source: "10.png"
    im_num: 10
    allow_stretch: True
    on_im_num: app.change_monst_im(self)
    on_pos_hint: app.monst_pos_hint(self)
```

```
<Character@Image>:
    size_hint: (0.15, 0.15)
    pos_hint: {'x': 0.2, 'y': 0.6}
    source: "0.png"
    im_num: 0
    allow_stretch: True
    on_im_num: app.change_char_im(self)
    on_pos_hint: app.char_pos_hint(self)
```

Python File

The Python file in Listing 6-22 includes the code for playing the sound effects, as we did in the previous single-level game.

Listing 6-22. Complete Python File of the Game

```python
import kivy.app
import kivy.uix.screenmanager
import random
import kivy.core.audio
import os
import functools

class TestApp(kivy.app.App):

    def screen_on_pre_leave(self, screen_num):
        curr_screen = self.root.screens[screen_num]

        curr_screen.ids['monster_image_lvl'+str(screen_num)].pos_hint =
        {'x': 0.8, 'y': 0.8}
        curr_screen.ids['character_image_lvl'+str(screen_num)].pos_hint =
        {'x': 0.2, 'y': 0.6}

    def screen_on_pre_enter(self, screen_num):
        curr_screen = self.root.screens[screen_num]
        curr_screen.character_killed = False
        curr_screen.num_coins_collected = 0
        curr_screen.ids['character_image_lvl'+str(screen_num)].im_num = 0
```

```
curr_screen.ids['monster_image_lvl'+str(screen_num)].im_num = 10
curr_screen.ids['num_coins_collected_lvl'+str(screen_num)].text =
"Coins 0"

for key, coin in curr_screen.coins_ids.items():
    curr_screen.ids['layout_lvl'+str(screen_num)].remove_widget(coin)
curr_screen.coins_ids = {}

coin_width = 0.05
coin_height = 0.05

curr_screen = self.root.screens[screen_num]

section_width = 1.0/curr_screen.num_coins
for k in range(curr_screen.num_coins):
    x = random.uniform(section_width*k, section_width*(k+1)-coin_
    width)
    y = random.uniform(0, 1-coin_height)
    coin = kivy.uix.image.Image(source="coin.png", size_hint=(coin_
    width, coin_height), pos_hint={'x': x, 'y': y}, allow_
    stretch=True)
    curr_screen.ids['layout_lvl'+str(screen_num)].add_widget(coin,
    index=-1)
    curr_screen.coins_ids['coin'+str(k)] = coin

def screen_on_enter(self, screen_num):
    music_dir = os.getcwd()+"/music/"
    self.bg_music = kivy.core.audio.SoundLoader.load(music_dir+"bg_
    music_piano.wav")
    self.bg_music.loop = True

    self.coin_sound = kivy.core.audio.SoundLoader.load(music_dir+"coin.
    wav")
    self.level_completed_sound = kivy.core.audio.SoundLoader.
    load(music_dir+"level_completed_flaute.wav")
    self.char_death_sound = kivy.core.audio.SoundLoader.load(music_
    dir+"char_death_flaute.wav")

    self.bg_music.play()
```

```python
    curr_screen = self.root.screens[screen_num]
    monster_image = curr_screen.ids['monster_image_lvl'+str(screen_num)]
    new_pos = (random.uniform(0.0, 1 - monster_image.size_hint[0]),
    random.uniform(0.0, 1 - monster_image.size_hint[1]))
    self.start_monst_animation(monster_image=monster_image, new_
    pos=new_pos, anim_duration=random.uniform(1.5, 3.5))

def start_monst_animation(self, monster_image, new_pos, anim_duration):
    monst_anim = kivy.animation.Animation(pos_hint={'x': new_pos[0],
    'y': new_pos[1]}, im_num=17,duration=anim_duration)
    monst_anim.bind(on_complete=self.monst_animation_completed)
    monst_anim.start(monster_image)

def monst_animation_completed(self, *args):
    monster_image = args[1]
    monster_image.im_num = 10

    new_pos = (random.uniform(0.0, 1 - monster_image.size_hint[0]),
    random.uniform(0.0, 1 - monster_image.size_hint[1]))
    self.start_monst_animation(monster_image=monster_image, new_pos=
    new_pos,anim_duration=random.uniform(1.5, 3.5))

def monst_pos_hint(self, monster_image):
    screen_num = int(monster_image.parent.parent.name[5:])
    curr_screen = self.root.screens[screen_num]
    character_image = curr_screen.ids['character_image_lvl'+str(screen_
    num)]

    character_center = character_image.center
    monster_center = monster_image.center

    gab_x = character_image.width / 2
    gab_y = character_image.height / 2
    if character_image.collide_widget(monster_image) and abs(character_
    center[0] - monster_center[0]) <= gab_x and abs(character_
    center[1] - monster_center[1]) <= gab_y and curr_screen.character_
    killed == False:
        self.bg_music.stop()
```

```python
            self.char_death_sound.play()
        curr_screen.character_killed = True

        kivy.animation.Animation.cancel_all(character_image)
        kivy.animation.Animation.cancel_all(monster_image)

        character_image.im_num = 91
        char_anim = kivy.animation.Animation(im_num=95, duration=1.0)
        char_anim.start(character_image)
        kivy.clock.Clock.schedule_once(functools.partial(self.back_to_
        main_screen, curr_screen.parent), 3)

    def change_monst_im(self, monster_image):
        monster_image.source = str(int(monster_image.im_num)) + ".png"

    def touch_down_handler(self, screen_num, args):
        curr_screen = self.root.screens[screen_num]
        if curr_screen.character_killed == False:
            self.start_char_animation(screen_num, args[1].spos)

    def start_char_animation(self, screen_num, touch_pos):
        curr_screen = self.root.screens[screen_num]
        character_image = curr_screen.ids['character_image_lvl'+str(screen_
        num)]
        character_image.im_num = 0
        char_anim = kivy.animation.Animation(pos_hint={'x': touch_pos[0] -
        character_image.size_hint[0] / 2,'y': touch_pos[1] - character_
        image.size_hint[1] / 2}, im_num=7)
        char_anim.bind(on_complete=self.char_animation_completed)
        char_anim.start(character_image)

    def char_animation_completed(self, *args):
        character_image = args[1]
        character_image.im_num = 0

    def char_pos_hint(self, character_image):
        screen_num = int(character_image.parent.parent.name[5:])
        character_center = character_image.center
```

```
    gab_x = character_image.width / 3
    gab_y = character_image.height / 3
    coins_to_delete = []
    curr_screen = self.root.screens[screen_num]

    for coin_key, curr_coin in curr_screen.coins_ids.items():
        curr_coin_center = curr_coin.center
        if character_image.collide_widget(curr_coin) and abs(character_
        center[0] - curr_coin_center[0]) <= gab_x and abs(character_
        center[1] - curr_coin_center[1]) <= gab_y:
            self.coin_sound.play()
            coins_to_delete.append(coin_key)
            curr_screen.ids['layout_lvl'+str(screen_num)].remove_
            widget(curr_coin)
            curr_screen.num_coins_collected = curr_screen.num_coins_
            collected + 1
            curr_screen.ids['num_coins_collected_lvl'+str(screen_num)].
            text = "Coins "+str(curr_screen.num_coins_collected)
            if curr_screen.num_coins_collected == curr_screen.num_coins:
                self.bg_music.stop()
                self.level_completed_sound.play()
                kivy.animation.Animation.cancel_all(character_image)
                kivy.clock.Clock.schedule_once(functools.partial(self.
                back_to_main_screen, curr_screen.parent), 3)
                kivy.animation.Animation.cancel_all(curr_screen.
                ids['monster_image_lvl'+str(screen_num)])

    if len(coins_to_delete) > 0:
        for coin_key in coins_to_delete:
            del curr_screen.coins_ids[coin_key]

def change_char_im(self, character_image):
    character_image.source = str(int(character_image.im_num)) + ".png"

def back_to_main_screen(self, screenManager, *args):
    screenManager.current = "main"
```

```python
class MainScreen(kivy.uix.screenmanager.Screen):
    pass

class Level1(kivy.uix.screenmanager.Screen):
    character_killed = False
    num_coins = 5
    num_coins_collected = 0
    coins_ids = {}

class Level2(kivy.uix.screenmanager.Screen):
    character_killed = False
    num_coins = 8
    num_coins_collected = 0
    coins_ids = {}

app = TestApp()
app.run()
```

Adding More Levels

Now that we have solved all the issues, we have successfully created a two-level game. We can easily add more levels to the game by following these steps:

1. Create a class for the level inside the Python file that extends the Screen class and initializes the previous four class variables.

2. Create a custom widget that defines the layout of the class inside the KV file. Remember to change the IDs of the three widgets added to the class (Monster, Character, and NumCoinsCollected) to have the screen index at the end.

3. Add an instance of the custom widget to the application widget tree as a child of the ScreenManager.

4. Do not forget to add a button inside the main screen in order to move to the new level.

By following these four steps, we can add more levels as needed.

Adding More Monsters

Each level can just have a single monster, which makes the game boring as there are no new challenges. What should we do in order to add more monsters with the least changes in the code? Let's see.

If we want to add more monsters, each one must be uniquely identified. This is because the location of each monster must be compared to the character location in case there is a collision. The way to differentiate between the different monsters is to give each one a different ID in the KV file. By referring to the ID of each monster, we can uniquely identify each monster.

The general form of the monster ID could be as follows:

monster**<monst_index>**_image_lvl**<lvl_index>**

Where monst_index refers to the monster index and lvl_index refers to the level index where each of them starts. For example, if each index is made to start at 1, the ID of the second monster in the third level will be monster2_image_lvl3. The layout of level 3, which has two monsters, is shown in Listing 6-23.

Listing 6-23. Defining the Layout for Level 3 in Which Two Monsters Exist

```
<Level3>:
    name: "level3"
    on_pre_enter: app.screen_on_pre_enter(3)
    on_pre_leave: app.screen_on_pre_leave(3)
    on_enter: app.screen_on_enter(3)
    FloatLayout:
        id: layout_lvl3
        on_touch_down: app.touch_down_handler(3, args)
        canvas.before:
            Rectangle:
                pos: self.pos
                size: self.size
                source: "bg_lvl3.jpg"
        NumCollectedCoins:
            id: num_coins_collected_lvl3
        Monster:
            id: monster1_image_lvl3
```

```
Monster:
    id: monster2_image_lvl3
Character:
    id: character_image_lvl3
```

Notice that we used the monster ID to refer to monster in the Python code inside each of the following callback functions:

- screen_on_pre_leave(): Reset the monster location

- screen_on_pre_enter(): Reset the im_num property

- screen_on_enter(): Start the animation

- monst_pos_hint(): Cancel its animation after the character is killed

- char_pos_hint(): Cancel its animation after the level is completed by collecting all the coins

The current form in which the ID is specified does not reflect the monster index. For that reason, we must change it.

Inside the char_pos_hint(), for example, the monster animation is cancelled according to this line:

```
kivy.animation.Animation.cancel_all(curr_screen.ids['monster_image_lvl' +
str(screen_num)])
```

When there is more than one monster, the previous line must be repeated for all of them. We can easily achieve that using a for loop that loops through all the monsters, prepares their IDs according to the monster index and the level (screen) index, and then uses this ID as a key to the ids dictionary.

In order to create a for loop, we need to define the number of monsters in the level. This is we add why a new class variable named num_monsters. The definition of the level 3 class (Level3) after adding this variable is shown in Listing 6-24.

Listing 6-24. Adding the num_monsters Variable Inside the Class Header of Level 3

```
class Level3(kivy.uix.screenmanager.Screen):
    character_killed = False
    num_coins = 12
    num_coins_collected = 0
```

```
coins_ids = {}
num_monsters = 2
```

Listing 6-25 shows the for loop added inside char_pos_hint() to access each monster and cancel its animation. The monster index is returned by (i + 1) because the loop variable i start at 0 and the first monster index is 1.

Listing 6-25. Accessing the Monsters and Cancelling Their Animations

```
for i in range(curr_screen.num_monsters): kivy.animation.Animation.cancel_
all(curr_screen.ids['monster' + str(i + 1) + '_image_lvl' + str(screen_num)])
```

The same way, we can edit the remaining four callback functions to access each monster and then apply the desired operation over it.

Game with Multiple Monsters per Level

The modified Python code of the game that supports using more than one monster per level is listed in Listing 6-26. Note that the num_monsters class variable must be added inside all levels classes. In the first two levels, it is set to 1. For the third class, it is 2. What if this variable is set to 0?

Listing 6-26. Python File for Supporting More Than One Monster Within the Game

```
import kivy.app
import kivy.uix.screenmanager
import random
import kivy.core.audio
import os
import functools

class TestApp(kivy.app.App):

    def screen_on_pre_leave(self, screen_num):
        curr_screen = self.root.screens[screen_num]
```

```
    for i in range(curr_screen.num_monsters):
        curr_screen.ids['monster'+str(i+1)+'_image_lvl'+str(screen_
        num)].pos_hint = {'x': 0.8, 'y': 0.8}
    curr_screen.ids['character_image_lvl'+str(screen_num)].pos_hint =
    {'x': 0.2, 'y': 0.6}

def screen_on_pre_enter(self, screen_num):
    curr_screen = self.root.screens[screen_num]
    curr_screen.character_killed = False
    curr_screen.num_coins_collected = 0
    curr_screen.ids['character_image_lvl'+str(screen_num)].im_num = 0
    for i in range(curr_screen.num_monsters):
        curr_screen.ids['monster'+str(i+1)+'_image_lvl'+str(screen_
        num)].im_num = 10
    curr_screen.ids['num_coins_collected_lvl'+str(screen_num)].text =
    "Coins 0"

    for key, coin in curr_screen.coins_ids.items():
        curr_screen.ids['layout_lvl'+str(screen_num)].remove_
        widget(coin)
    curr_screen.coins_ids = {}

    coin_width = 0.05
    coin_height = 0.05

    curr_screen = self.root.screens[screen_num]

    section_width = 1.0/curr_screen.num_coins
    for k in range(curr_screen.num_coins):
        x = random.uniform(section_width*k, section_width*(k+1)-coin_
        width)
        y = random.uniform(0, 1-coin_height)
        coin = kivy.uix.image.Image(source="coin.png", size_hint=(coin_
        width, coin_height), pos_hint={'x': x, 'y': y}, allow_
        stretch=True)
        curr_screen.ids['layout_lvl'+str(screen_num)].add_widget(coin,
        index=-1)
        curr_screen.coins_ids['coin'+str(k)] = coin
```

```python
def screen_on_enter(self, screen_num):
    music_dir = os.getcwd()+"/music/"
    self.bg_music = kivy.core.audio.SoundLoader.load(music_dir+"bg_
    music_piano.wav")
    self.bg_music.loop = True

    self.coin_sound = kivy.core.audio.SoundLoader.load(music_dir+"coin.
    wav")
    self.level_completed_sound = kivy.core.audio.SoundLoader.
    load(music_dir+"level_completed_flaute.wav")
    self.char_death_sound = kivy.core.audio.SoundLoader.load(music_
    dir+"char_death_flaute.wav")

    self.bg_music.play()

    curr_screen = self.root.screens[screen_num]
    for i in range(curr_screen.num_monsters):
        monster_image = curr_screen.ids['monster'+str(i+1)+'_image_
        lvl'+str(screen_num)]
        new_pos = (random.uniform(0.0, 1 - monster_image.size_
        hint[0]/4), random.uniform(0.0, 1 - monster_image.size_
        hint[1]/4))
        self.start_monst_animation(monster_image=monster_image, new_
        pos=new_pos, anim_duration=random.uniform(1.0, 3.0))
    new_pos = (random.uniform(0.0, 1 - monster_image.size_hint[0]),
    random.uniform(0.0, 1 - monster_image.size_hint[1]))
    self.start_monst_animation(monster_image=monster_image, new_
    pos=new_pos, anim_duration=random.uniform(1.5, 3.5))

def start_monst_animation(self, monster_image, new_pos, anim_duration):
    monst_anim = kivy.animation.Animation(pos_hint={'x': new_pos[0],
    'y': new_pos[1]}, im_num=17,duration=anim_duration)
    monst_anim.bind(on_complete=self.monst_animation_completed)
    monst_anim.start(monster_image)

def monst_animation_completed(self, *args):
    monster_image = args[1]
    monster_image.im_num = 10
```

```
        new_pos = (random.uniform(0.0, 1 - monster_image.size_hint[0]),
        random.uniform(0.0, 1 - monster_image.size_hint[1]))
        self.start_monst_animation(monster_image=monster_image, new_pos=
        new_pos,anim_duration=random.uniform(1.5, 3.5))

    def monst_pos_hint(self, monster_image):
        screen_num = int(monster_image.parent.parent.name[5:])
        curr_screen = self.root.screens[screen_num]
        character_image = curr_screen.ids['character_image_lvl'+str(screen_
        num)]

        character_center = character_image.center
        monster_center = monster_image.center

        gab_x = character_image.width / 2
        gab_y = character_image.height / 2
        if character_image.collide_widget(monster_image) and abs(character_
        center[0] - monster_center[0]) <= gab_x and abs(character_
        center[1] - monster_center[1]) <= gab_y and curr_screen.character_
        killed == False:
            self.bg_music.stop()
            self.char_death_sound.play()
            curr_screen.character_killed = True

            kivy.animation.Animation.cancel_all(character_image)
            for i in range(curr_screen.num_monsters):
                kivy.animation.Animation.cancel_all(curr_screen.
                ids['monster'+str(i+1)+'_image_lvl'+str(screen_num)])

            character_image.im_num = 91
            char_anim = kivy.animation.Animation(im_num=95, duration=1.0)
            char_anim.start(character_image)
            kivy.clock.Clock.schedule_once(functools.partial(self.back_to_
            main_screen, curr_screen.parent), 3)

    def change_monst_im(self, monster_image):
        monster_image.source = str(int(monster_image.im_num)) + ".png"
```

```python
def touch_down_handler(self, screen_num, args):
    curr_screen = self.root.screens[screen_num]
    if curr_screen.character_killed == False:
        self.start_char_animation(screen_num, args[1].spos)

def start_char_animation(self, screen_num, touch_pos):
    curr_screen = self.root.screens[screen_num]
    character_image = curr_screen.ids['character_image_lvl'+str(screen_num)]
    character_image.im_num = 0
    char_anim = kivy.animation.Animation(pos_hint={'x': touch_pos[0] -
    character_image.size_hint[0] / 2,'y': touch_pos[1] - character_
    image.size_hint[1] / 2}, im_num=7)
    char_anim.bind(on_complete=self.char_animation_completed)
    char_anim.start(character_image)

def char_animation_completed(self, *args):
    character_image = args[1]
    character_image.im_num = 0

def char_pos_hint(self, character_image):
    screen_num = int(character_image.parent.parent.name[5:])
    character_center = character_image.center

    gab_x = character_image.width / 3
    gab_y = character_image.height / 3
    coins_to_delete = []
    curr_screen = self.root.screens[screen_num]

    for coin_key, curr_coin in curr_screen.coins_ids.items():
        curr_coin_center = curr_coin.center
        if character_image.collide_widget(curr_coin) and abs(character_
        center[0] - curr_coin_center[0]) <= gab_x and abs(character_
        center[1] - curr_coin_center[1]) <= gab_y:
            self.coin_sound.play()
            coins_to_delete.append(coin_key)
            curr_screen.ids['layout_lvl'+str(screen_num)].remove_
            widget(curr_coin)
```

```
            curr_screen.num_coins_collected = curr_screen.num_coins_
            collected + 1
            curr_screen.ids['num_coins_collected_lvl'+str(screen_num)].
            text = "Coins "+str(curr_screen.num_coins_collected)
            if curr_screen.num_coins_collected == curr_screen.num_
            coins:
                self.bg_music.stop()
                self.level_completed_sound.play()
                kivy.animation.Animation.cancel_all(character_image)
                kivy.clock.Clock.schedule_once(functools.partial(self.
                back_to_main_screen, curr_screen.parent), 3)
                for i in range(curr_screen.num_monsters):
                    kivy.animation.Animation.cancel_all(curr_screen.
                    ids['monster' + str(i + 1) + '_image_lvl' +
                    str(screen_num)])

        if len(coins_to_delete) > 0:
            for coin_key in coins_to_delete:
                del curr_screen.coins_ids[coin_key]

    def change_char_im(self, character_image):
        character_image.source = str(int(character_image.im_num)) + ".png"

    def back_to_main_screen(self, screenManager, *args):
        screenManager.current = "main"

class MainScreen(kivy.uix.screenmanager.Screen):
    pass

class Level1(kivy.uix.screenmanager.Screen):
    character_killed = False
    num_coins = 5
    num_coins_collected = 0
    coins_ids = {}
    num_monsters = 1

class Level2(kivy.uix.screenmanager.Screen):
    character_killed = False
```

```
    num_coins = 8
    num_coins_collected = 0
    coins_ids = {}
    num_monsters = 1

class Level3(kivy.uix.screenmanager.Screen):
    character_killed = False
    num_coins = 12
    num_coins_collected = 0
    coins_ids = {}
    num_monsters = 2

app = TestApp()
app.run()
```

If the num_monsters variable is set to 0, no for loop will be executed, because the character will not animate and thus its position won't change. The on_pos_hint event will not be fired. Because collision detection takes place inside the callback function of that event, no collision will occur between the monster and the character. The monster will be visible because it is added inside the KV file. Listing 6-27 shows the modified KV file after creating three levels.

Listing 6-27. Adding Three Levels Inside the KV File

```
ScreenManager:
    MainScreen:
    Level1:
    Level2:
    Level3:

<MainScreen>:
    name: "main"
    BoxLayout:
        Button:
            text: "Go to Level 1"
            on_press: app.root.current="level1"
        Button:
            text: "Go to Level 2"
```

```
                on_press: app.root.current="level2"
        Button:
            text: "Go to Level 3"
            on_press: app.root.current = "level3"

<Level1>:
    name: "level1"
    on_pre_enter: app.screen_on_pre_enter(1)
    on_pre_leave: app.screen_on_pre_leave(1)
    on_enter: app.screen_on_enter(1)
    FloatLayout:
        id: layout_lvl1
        on_touch_down: app.touch_down_handler(1, args)
        canvas.before:
            Rectangle:
                size: self.size
                pos: self.pos
                source: "bg_lvl1.jpg"
        NumCollectedCoins:
            id: num_coins_collected_lvl1
        Monster:
            id: monster_image_lvl1
        Character:
            id: character_image_lvl1

<Level2>:
    name: "level2"
    on_pre_enter: app.screen_on_pre_enter(2)
    on_pre_leave: app.screen_on_pre_leave(2)
    on_enter: app.screen_on_enter(2)
    FloatLayout:
        id: layout_lvl2
        on_touch_down: app.touch_down_handler(2, args)
        canvas.before:
            Rectangle:
                size: self.size
```

```
                pos: self.pos
                source: "bg_lvl2.jpg"
        NumCollectedCoins:
            id: num_coins_collected_lvl2
        Monster:
            id: monster_image_lvl2
        Character:
            id: character_image_lvl2

<Level3>:
    name: "level3"
    on_pre_enter: app.screen_on_pre_enter(3)
    on_pre_leave: app.screen_on_pre_leave(3)
    on_enter: app.screen_on_enter(3)
    FloatLayout:
        id: layout_lvl3
        on_touch_down: app.touch_down_handler(3, args)
        canvas.before:
            Rectangle:
                size: self.size
                pos: self.pos
                source: "bg_lvl3.jpg"
        NumCollectedCoins:
            id: num_coins_collected_lvl3
        Monster:
            id: monster1_image_lvl3
        Monster:
            id: monster2_image_lvl3
        Character:
            id: character_image_lvl3

<NumCollectedCoins@Label>:
    size_hint: (0.1, 0.02)
    pos_hint: {'x': 0.0, 'y': 0.97}
    text: "Coins 0"
    font_size: 20
```

```
<Monster@Image>:
    size_hint: (0.15, 0.15)
    pos_hint: {'x': 0.8, 'y': 0.8}
    source: "10.png"
    im_num: 10
    allow_stretch: True
    on_im_num: app.change_monst_im(self)
    on_pos_hint: app.monst_pos_hint(self)

<Character@Image>:
    size_hint: (0.15, 0.15)
    pos_hint: {'x': 0.2, 'y': 0.6}
    source: "0.png"
    im_num: 0
    allow_stretch: True
    on_im_num: app.change_char_im(self)
    on_pos_hint: app.char_pos_hint(self)
```

Tips About Widgets Properties

Each widget inside each level will have a number of properties and you should
determine where to define them. Here are some tips:

- The properties shared across all widgets inside the class should be
  added to the class header in the Python file.

- The properties shared across all instances from a specific widget
  but not shared across other types of widgets should be added to the
  definition of the custom widget in the KV file.

- The properties that change from one instance from one widget
  to another instance of the same widget should be added to the
  definition of the instance of the custom widget in the KV file.

These rules are useful to avoid repeating some parts of the code. For example, say
we want to define a property that defines the animation duration across all monsters in
all levels. Because that property is static across all monsters and across all levels, we can

create a single variable inside the application class to do the job. If the variable is added within the levels Screen classes, the same variable will be repeated for each class, which is not required.

If it is possible to assign different animation duration for monsters in different levels, that variable cannot be defined in the application class and it's better to add it to the Screen class headers.

If the monster duration animation is to be changed for each monster even within the same level, we need to assign a property to each instance of the Monster widget.

The next section changes both the character and monster animations to apply this understanding.

Changing Animation Durations

In the previous game, the animation duration of the character and all monsters are fixed across all levels. The duration can be changed in order to make the game more challenging. For example, the player would have to make fast decisions if the monsters moved more quickly. In order to change the duration, we have to ask ourselves where the properties that define the duration should be specified.

Regarding the character, we want to change its duration in each level. Because it exists only once in the level, the character duration should be specified inside the Python level class. Listing 6-28 shows the Level3 class header after adding a property called char_anim_duration.

Listing 6-28. Adding the char_anim_duration Property to the Class of Level 3

```
class Level3(kivy.uix.screenmanager.Screen):
    character_killed = False
    num_coins = 12
    num_coins_collected = 0
    coins_ids = {}
    char_anim_duration = 1.2
    num_monsters = 2
```

Regarding the monsters, we also want to change their duration in each level. But because there is more than one monster per level, we cannot specify their duration inside the class header, but instead inside the KV widget instance. Listing 6-29 specifies the duration of the two monsters in level 3 inside the KV file. Because the monster animation is generated randomly by setting the minimum and maximum possible values, the properties monst_anim_duration_low and monst_anim_duration_high set the minimum and maximum values of the duration.

Listing 6-29. Specifying the Monster Duration Within the KV File

```
<Level3>:
    name: "level3"
    on_pre_enter: app.screen_on_pre_enter(3)
    on_pre_leave: app.screen_on_pre_leave(3)
    on_enter: app.screen_on_enter(3)
    FloatLayout:
        id: layout_lvl3
        on_touch_down: app.touch_down_handler(3, args)
        canvas.before:
            Rectangle:
                pos: self.pos
                size: self.size
                source: "levels-bg/bg_lvl3.jpg"
        NumCollectedCoins:
            id: num_coins_collected_lvl3
        Monster:
            id: monster1_image_lvl3
            monst_anim_duration_low: 1.0
            monst_anim_duration_high: 1.6
        Monster:
            id: monster2_image_lvl3
            monst_anim_duration_low: 1.0
            monst_anim_duration_high: 2.0
        Character:
            id: character_image_lvl3
```

After setting the duration for both, we need to reference them in the Python code. The character animation is referenced one time inside the start_char_animation() callback function. Listing 6-30 shows the modified function that references the char_anim_duration variable defined in the class.

Listing 6-30. Referencing the char_anim_duration Variable to Return the Duration of the Monster Character

```
def start_char_animation(self, screen_num, touch_pos):
    curr_screen = self.root.screens[screen_num]
    character_image = curr_screen.ids['character_image_lvl' + str(screen_
    num)]
    character_image.im_num = character_image.start_im_num
    char_anim = kivy.animation.Animation(pos_hint={'x': touch_pos[0] -
    character_image.size_hint[0] / 2, 'y': touch_pos[1] - character_image.
    size_hint[1] / 2},im_num=character_image.end_im_num, duration=curr_
    screen.char_anim_duration)
    char_anim.bind(on_complete=self.char_animation_completed)
    char_anim.start(character_image)
```

Regarding the monster animation, it is referenced two times. The first reference is inside the screen_on_enter() callback function in order to start the animation as soon as the screen starts. It is listed in Listing 6-31 after being modified to use monst_anim_duration_low and monst_anim_duration_high defined inside the Monster widget.

Listing 6-31. Setting the Monster Animation Duration Inside the screen_on_enter() Callback Function

```
def screen_on_enter(self, screen_num):
    music_dir = os.getcwd() + "/music/"
    self.bg_music = kivy.core.audio.SoundLoader.load(music_dir + "bg_music_
    piano.wav")
    self.bg_music.loop = True

    self.coin_sound = kivy.core.audio.SoundLoader.load(music_dir + "coin.
    wav")
    self.level_completed_sound = kivy.core.audio.SoundLoader.load(music_dir
    + "level_completed_flaute.wav")
    self.char_death_sound = kivy.core.audio.SoundLoader.load(music_dir +
    "char_death_flaute.wav")

    self.bg_music.play()
```

```
curr_screen = self.root.screens[screen_num]
for i in range(curr_screen.num_monsters):
    monster_image = curr_screen.ids['monster' + str(i + 1) + '_image_
    lvl' + str(screen_num)]
    new_pos = (random.uniform(0.0, 1 - monster_image.size_hint[0] / 4),
               random.uniform(0.0, 1 - monster_image.size_hint[1] / 4))
    self.start_monst_animation(monster_image=monster_image, new_
                               pos=new_pos, anim_duration=random.
                               uniform(monster_image.monst_anim_
                               duration_low, monster_image.monst_anim_
                               duration_high))

for i in range(curr_screen.num_fires):
    fire_widget = curr_screen.ids['fire' + str(i + 1) + '_lvl' +
    str(screen_num)]
    self.start_fire_animation(fire_widget=fire_widget, pos=(0.0, 0.5),
    anim_duration=5.0)
```

The second time the monster animation is referenced is inside the callback function called monst_animation_completed(), which is listed in Listing 6-32. It repeats the animation after it's complete.

Listing 6-32. Setting the Monster Animation Duration Inside the monst_animation_completed() Callback Function

```
def monst_animation_completed(self, *args):
    monster_image = args[1]
    monster_image.im_num = monster_image.start_im_num

    new_pos = (
    random.uniform(0.0, 1 - monster_image.size_hint[0] / 4), random.
    uniform(0.0, 1 - monster_image.size_hint[1] / 4))
    self.start_monst_animation(monster_image=monster_image, new_pos=new_pos,
                               anim_duration=random.uniform(monster_image.
                               monst_anim_duration_low, monster_image.
                               monst_anim_duration_high))
```

More To Do

Note that there is more you can do to the game to make it more interesting. It is up to your imagination to suggest new ideas that make the game more interesting. Consider these changes to the previous game as examples:

- The `NumCollectedCoins` label can show the number of collected coins in addition to the total number of coins per level. This helps players know how many coins are left to collect.

- The new label can be added to display the level number.

- In addition to the monsters, fire can be thrown in order to kill the character.

- Rather than killing the character in the first collision, we could wait for two or three collisions.

- The character also may kill the monsters using fire.

- Rather than creating new monsters that use the same images, a new `Monster` widget could be added with different images. The new properties `start_im_num` and `end_im_num` can be added to specify the values of the first and last values of the `im_num` property. The properties are added inside the custom widget, not in their instances, because they are identical in all monsters and characters. This makes creating new monsters easier.

- Limit the space that the character can move per touch. It is unlimited currently but we can limit the character to move a quarter of the screen's width and height per touch.

- Change the random range of `pos_hint` for the monster animation to be near the current position of the character.

- Ask the player to collect a certain number of coins in a limited period.

- Add bonus levels.

- Create a bonus for the player when five coins are collected in a single touch.

- Open level numbered i only when completing level i-1 by saving the number of the highest level completed in a file. This number is updated after each level is completed and restored upon the application start.

- Rate the completed level according to the number of collisions. For example, two starts for no collisions, two stars for 10 collisions, and three starts for more than 10 collisions.

Some of these changes are applied to the game according to the KV and Python files shown in Appendix.

Previously, a button referred to each level inside the main screen. This button is replaced with a new custom widget named ImageButton, which is a hybrid of the Button and Image widgets. These new widgets have the source property to add an image and also handle the on_press and on_release events.

The ImageButton widgets referring to the levels are added in GridLayout. This arranges the widgets in a grid rather than in a box, which helps to add more widgets to the screen. A background image is added to the GridLayout, as shown in Figure 6-14.

A new button at the top of the main screen is added and it navigates the application to a new screen given class name AboutUs, which prints details about the developer. Remember to add this new screen as a child inside the ScreenManager to not affect the indices of the levels.

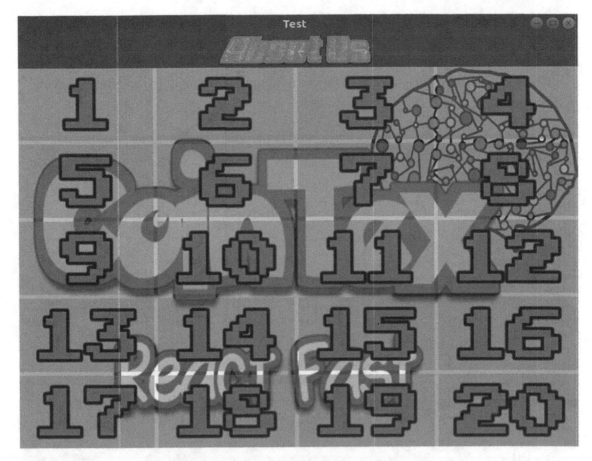

Figure 6-14. *Adding a background image to the main screen*

A new widget called Fire that extends the Label widget is created to represent thrown fire that kills the player. Using canvas.before, a fire image background is added using the Rectangle vertex instruction. Inside the instance of the widget, two properties are added called fire_start_pos_hint and fire_end_pos_hint. They specify the path in which the fire follows. Because there may be more than one Fire widget in the same level, they are given IDs similar to how the monsters are given their IDs. For example, the first and second Fire widgets in the third level have the fire1_lvl3 and fire2_lvl3 IDs, respectively.

Similar to the num_monsters class variable, there is a variable named num_fires that holds the number of Fire widgets inside the level. When an operation is to be applied on the Fires, a loop iterates over all of them.

The way to handle the position and collision of the new Fire widget is very similar to the Monster widget. The on_pos_hint event is attached to the Fire widget, which

285

is handled using the callback function `fire_pos_hint()`. Inside that function, the fire position is compared to the monster position. When collision occurs, the same stuff inside the `most_pos_hint()` function is repeated.

The screen of level 20 is shown in Figure 6-15, and it has eight `Fire` widgets moving in different directions.

Figure 6-15. *The screen of level 20, which has eight Fire widgets*

According to Figure 6-15, a label is added next to the `NumCollectedCoins` to display the current level number. That widget is named `LevelNumber` inside the KV file.

In the new game, the character is no longer killed in the first collision. A maximum number of collisions is specified inside the `num_collision_level` class variable inside each level. Another class variable named `num_collisions_hit` starts at 0 and is incremented each time a collision occurs. When the values in `num_collision_level` and `num_collisions_hit` are equal, the character will die.

There is also a red bar next to the level number, which represents the remaining collisions of the character. This bar is created using the Rectangle and Color vertex instructions inside a custom widget called RemainingLifePercent, which extends the Label widget. For each collision, the size of that widget is reduced inside the callback functions that are responsible for collision detection, which are monst_pos_hint() and fire_pos_hint().

A new monster widget called Monster2 is created and it behaves almost identically to the Monster widget. The difference is that it uses new values for the im_num property. For that reason, the new properties start_im_num and end_im_num accept the first and last numbers of the image numbers. Inside the Python code, the values in such properties are used in the animations. Two instances of Monster2 are used in level 16, according to the screen displayed in Figure 6-16.

Figure 6-16. *Two different monsters used in level 16*

These two properties are also used in the Character widget to make them all act the same way. The character also has dead_start_im_num and dead_end_im_num properties, which refer to the numbers of the displayed images when the character is killed.

It is important to provide the correct paths of these files (images and audio). The new game organizes the images into these three folders:

- levels-bg: Holds the background images of the levels where the image name is bg_lvl<num>.jpg after replacing <num> with the level number. Remember that the level number starts at 1.

- levels-images: Holds the background images used for the ImageButton widget displayed on the main screen where the image name is <num>.png after replacing <num> with the level number.

- other-images: Holds the image displayed on the ImageButton widget referring to the AboutUs screen with a background image named About-Us.png. Inside that screen, there is another ImageButton that refers to the main screen with a background image named Main-Screen.png. The About Us screen is shown in Figure 6-17.

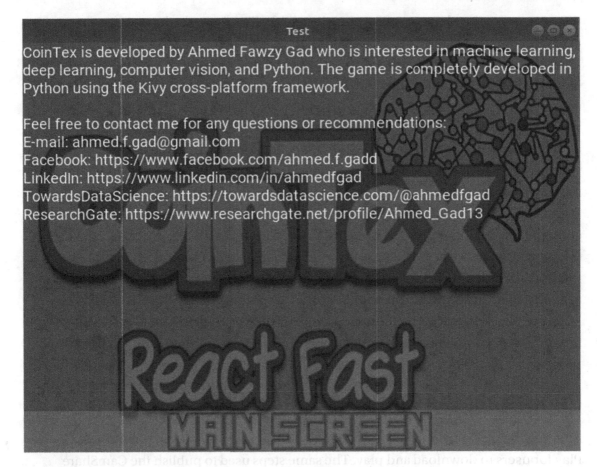

Figure 6-17. *About Us screen*

All audio files are inside the music folder with the following names:

- bg_music_piano_flute.wav: Background music of the main screen.

- bg_music_piano.wav: Background music of each level.

- char_death_flaute.wav: Played after the character is killed.

- coin.wav: Played when a coin is collected.

- level_completed_flaute.wav: Played after completing a level.

You can create the background sequences for free using https://onlinesequencer.net.

The game also opens level i only if level i-1 is complete. The idea is to enable all ImageButton widgets of all completed levels inside the main screen. This is in addition to enabling the ImageButton of a new level to be played. This requires knowing the last level completed by the player when the game starts. In order to do that, the game saves the progress of the player in a file using the `pickle` library. To do that, two new methods are created—`read_game_info()` and `activate_levels()`.

The `read_game_info()` method reads a file named `game_info`, which stores a dictionary with two items. The first item has a key `lastlvl` representing the last level completed by the player. The second item has a key of `congrats_displayed_once`. This items helps to display a congratulations message to the player once after completing all levels. The value of this item is `False` by default, meaning that the message is not yet displayed. It changes to `True` when all levels are completed and a congratulations message is displayed.

Inside the `on_start()` method, the `read_game_info()` method is called and the content (i.e., dictionary) stored in the dictionary is returned using the `pickle.load()` function. The dictionary values are returned and then fed as arguments to the second method `activate_levels()`. This method loops through all `ImageButton` widgets and activates them according to the number of last level completed.

Note that the `game_info` file is updated inside the `char_pos_hint()` callback function after completing each level using the `pickle.dump()` function.

Publishing the Game at Google Play

After testing the game and making sure it works as expected, we can publish it at Google Play for users to download and play. The same steps used to publish the CamShare application are repeated for this game.

After following the APK signing process, a release APK is created using the following command. Remember to set the minimum API of the project to at least 26.

```
ahmedgad@ubuntu:~/Desktop$ buildozer android release
```

If the game is titled `CoinTex`, it will be available at Google Play, as shown in Figure 6-18.

CoinTex is a multi-level adventure game with the mission of collecting all coins which are randomly distributed. Monsters and thrown fire struggle the player from completing the mission. Their motion are not expected and thus it tests the player's ability to do fast reactions in order to avoid their collision.

Figure 6-18. *The Android version of the CoinTex game is available at Google Play*

Complete Game

The complete source code of CoinTex is available in the appendix.

Summary

In summary, this chapter continued developing the game that we started in Chapter 5 by adding more levels. The game interface was organized using screens. More monsters were added. While developing the game, we covered issues that came up and discussed their solutions. We published the game for Android devices at Google Play under the name CoinTex for any user to download and install. By finishing this game, you will have a solid understanding of many of the Kivy features.

Python is a good way for developers to build Android applications but it cannot build rich applications like the ones developed in using Java Android Studio, which is the official IDE for building Android applications. The next chapter introduces using Android Studio for enriching the Python applications created in Kivy. If a feature is not supported by Python, do not worry, because it can be added in Android Studio.

## CHAPTER 7

Working with Kivy Projects from Android Studio

When a Python developer knows that it is possible to create Android apps using Python, one of the first questions to be asked is whether Python can develop rich Android applications like the ones developed in Android Studio using native Android languages (i.e. Java). Unfortunately, the answer is that Python is limited in its capabilities compared to Java when building Android (mobile) applications. But the good news is that there are ways to enrich the Android applications created in Python.

The Kivy team has developed some Python libraries that allow developers to access the Android features in a much easier way than was done in Android Studio. With just a few lines of Python code, we can access an Android feature. One of these libraries is Plyer, and it's discussed in this chapter. Using just a single line, a notification can be pushed to the Android notification bar.

Because Plyer is still under development, some of its features are unimplemented. One alternate solution is to reflect Java classes within Python using a library called Pyjnius. That involves writing Java code inside the Python file. This helps us access some of the Android features, but still there are some unresolved issues. Reflecting the Java classes within Python can increase the complexity of developing the application. Moreover, some exceptions occur when reflecting some Java classes. As a result, using Plyer and Pyjnius is not enough to access all the Android features.

Remember, as mentioned in Chapter 1, an Android Studio project is created when we build the Kivy application using Buildozer. This project could be imported easily within Android Studio and then we can continue developing the application there. This enables us to build whatever features we want.

293

© Ahmed Fawzy Mohamed Gad 2019
A. F. M. Gad, *Building Android Apps in Python Using Kivy with Android Studio*,
https://doi.org/10.1007/978-1-4842-5031-0_7

In Kivy, there is a problem packaging some Python libraries, such as OpenCV. If we cannot build an Android application from a Kivy application that uses OpenCV, we can avoid using OpenCV in Python and then use it in Android Studio. This workaround will be illustrated at the end of this chapter.

Plyer

In the applications created in the previous chapters, we used two main Android features that Kivy supports—accessing the camera and playing audio. Everything is done in Python. But there are other features in Android that might make life easier and we cannot access then in Kivy. We may use the gyroscope in the CoinTex game to move the main character. For CamShare, we might push notifications to the notification bar when the connection between the client and the server is lost. But how do we do that in Kivy (i.e., in Python)? Unfortunately, Kivy alone cannot access these features.

The Kivy team created a library called Pyjnius to access Java classes in Python. So, if you cannot access a feature from Python, you can access it from Java. Note that the Java code is written within the Python file and thus the code will not be Pythonic.

To solve that issue, the team created a library called `Plyer` to access the Android features inside the Python code. It is yet underdevelopment and some features are not available in the current version (1.3.3.dev0), such as the camera, audio, and Wi-Fi. The interface is very easy to learn and it accesses these features with minimal lines of Python. The Plyer documentation is available at `https://plyer.readthedocs.io/en/latest`.

In this section, a simple example is discussed in which Plyer is used to push notifications to the notification bar and change the orientation. In the applications created in the previous chapters, we tested them in a desktop computer before building the Android application. In this chapter, we cannot test the application in a desktop before building the Android application because the libraries used (e.g., Plyer) are running in Android devices. It is preferred to use logcat for monitoring the exceptions if any appear when running the Android application.

Changing Orientation and Pushing Notifications

Using the `plyer.notification`, we can push notifications to the notification bar of the Android device. This module has a function named `notify` that accepts a number of arguments to initialize the notification.

In order to change the device orientation, the plyer.orientation exists. It has three functions as follows:

- set_landscape(reverse=False): Sets the orientation to landscape if the reverse argument is True.

- set_portrait(reverse=False): Sets the orientation to portrait if the reverse argument is True.

- set_sensor(mode='any|landscape|portrait'): Sets the orientation according to the value specified in the mode argument.

Listing 7-1 shows the KV file of an application with three button widgets. The first one executes the show_notification() function, which pushes the notification to the notification bar.

When the second button is pressed, the orientation changes to landscape. Note that plyer.orientation is imported inside the KV file. The set_landscape() function is called within the KV file. The third button works like the second one, but calls the set_portrait() function to change the orientation to portrait.

Listing 7-1. KV File with Buttons for Pushing a Notification Message and Changing the Orientation of the Screen

```
#:import orientation plyer.orientation
BoxLayout:
    orientation: "vertical"
    Button:
        text: "Show Notification"
        on_press: app.show_notification()
    Button:
        text: "Portrait"
        on_press: orientation.set_portrait(reverse=True)
    Button:
        text: "Landscape"
        on_press: orientation.set_landscape(reverse=True)
```

The Python file is shown in Listing 7-2. The show_notification() function has a single line that calls the notify() function. The title and message arguments appear on the notification bar.

Listing 7-2. Pushing a Notification Message to the Android Notification Bar

```
import kivy.app
import plyer

class PushNotificationApp(kivy.app.App):

    def show_notification(self):
        plyer.notification.notify(title='Test', message='Notification using
        Plyer')

app = PycamApp()
app.run()
```

We used to run the application in a desktop computer for testing before building the Android application. Because Plyer is an Android package, it cannot be tested on a desktop computer and must run directly on Android devices. In order to make the debugging process easier, we use logcat for tracing the error stack. Remember that logcat can be activated while building, deploying, and running the APK file using the following command:

```
ahmedgad@ubuntu:~/Desktop$ buildozer android debug deploy run logcat
```

After deploying and running the application on an Android device and then pressing the first button, we will see the notification shown in Figure 7-1. The application icon is used as the icon of the notification. This is the least amount of code required to push a notification.

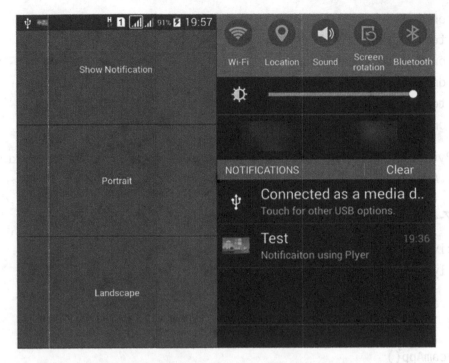

Figure 7-1. *An Android notification message created in Python using Plyer*

Controlling the Android Flashlight

The Android flashlight can be accessed in Plyer using `plyer.flash`. The KV file in Listing 7-3 creates three button widgets for turning on, turning off, and releasing the flash. The functions used for doing such work are listed here:

- `plyer.flash.on()`: Turns on the flash

- `plyer.flash.off()`: Turns off the flash

- `plyer.flash.release()`: Releases the flash

Listing 7-3. Controlling the Android Flashlight Using Plyer in Python

```
#:import flash plyer.flash
BoxLayout:
    orientation: "vertical"
    Button:
        text: "Turn On"
        on_press: flash.on()
```

297

```
Button:
    text: "Turn Off"
    on_press: flash.off()
Button:
    text: "Release"
    on_press: flash.release()
```

The Python file in Listing 7-4 does nothing except create a new class that extends the `kivy.app.App` class.

Listing 7-4. Generic Python File for Developing a Kivy Application

```python
import kivy.app
import plyer

class PycamApp(kivy.app.App):
    pass

app = PycamApp()
app.run()
```

After running the application, we will see the window shown in Figure 7-2. Remember to list `CAMERA` and `FLASHLIGHT` as items in the `android.permissions` property of the `buildozer.spec` file.

You can extend this application to create flash patterns that turn on and off the flashlight in a specified period of time.

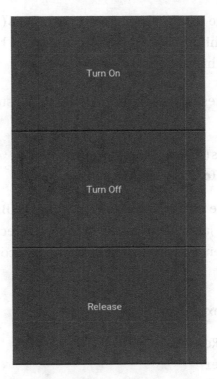

Figure 7-2. *An Android application for controlling the flashlight*

The previous examples show how simple the Plyer library is for accessing Android features within Python code. Unfortunately, Plyer is restricted to some features and does not do everything an Android developer wants to do. There are some unsupported features, such as displaying a toast message, and there are some features that are not implemented for some platforms, such as accessing the camera and playing audio. We can use the Pyjnius library to access these missing features using Java code.

Pyjnius

Pyjnius is a library developed by the Kivy team to access Java classes in the Python code in order to use the currently unsupported Android features within Python Kivy projects. The library has a core function named `autoclass()` that accepts the Java class name and returns a variable that represents that Java class. This process is called *reflection*.

A very simple example is shown in Listing 7-5. It prints a message using the `println()` method, which is in the `java.lang.System` class. The first statement imports the Pyjnius library. Then the Java class is reflected in Python by passing its name prepended with the package name as input to the `autoclass()` function. The returned variable named `System` in this example represents that class, but in Python. We can access the methods inside this class, as we have done in Java.

Listing 7-5. Printing a Message in Python Using the System Class in Java

```
import jnius
System = jnius.autoclass("java.lang.System")
System.out.println("Hello Java within Python")
```

Another simple example reads the contents of a text file. Before reading the file, we must reflect all required Java classes inside Python. It is recommended that we first prepare the Java file and then convert it to Python. The Java code needed to read a text file is shown in Listing 7-6.

Listing 7-6. Reading a Text File in Java

```
import java.io.BufferedReader
import java.io.FileReader

FileReader textFile = new FileReader("pycam.kv");
BufferedReader br = new BufferedReader(textFile);

StringBuilder sb = new StringBuilder();
String line = br.readLine();

while (line != null) {
    System.out.println(line);
    line = br.readLine();
}
```

There are three classes used in the Java code in Listing 7-6. The first one is the `java.io.FileReader` class for reading the file. In order to read lines inside the read file, the `java.io.BufferedReader` class is used. Finally, we can print each line on the console using the `java.lang.System` class. Note that the `java.lang` package is already imported in any Java class and thus we don't need to add an `import` statement for it within the Java code. After preparing the Java code, we need to embed it into Python using Pyjnius.

The previous three classes in Listing 7-6 are reflected in Python, as shown in Listing 7-7, using the `jnius.autoclass()` function. Then the filename is passed as input to the constructor of the FileReader class and the result returned into the textFile variable. In this example, the KV file previously created in Listing 7-3 for controlling the flashlight in Android is read. The file object is returned in the textFile variable, which is passed as input to the constructor of the BufferedReader class. The result is returned in the bufferedReader variable.

Listing 7-7. Reading a Text File Using Java Classes Reflected in Python Using Pyjnius

```python
import jnius

FileReader = jnius.autoclass("java.io.FileReader")
BufferedReader = jnius.autoclass("java.io.BufferedReader")
System = jnius.autoclass("java.lang.System")

textFile = FileReader("pycam.kv")

bufferedReader = BufferedReader(text_file)
line = bufferedReader.readLine()

while line != None:
    System.out.println(line)
    line = bufferedReader.readLine()
```

In order to read a line from the file, we call the readLine() Java method. To read all lines inside the file, a while loop is created for printing the lines as long as the returned line is not None. Note that None refers to null in Java. The lines printed to the console are shown in Figure 7-3.

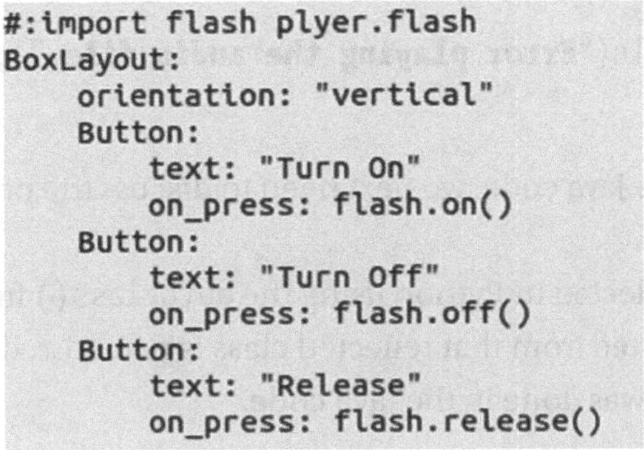

```
#:import flash plyer.flash
BoxLayout:
    orientation: "vertical"
    Button:
        text: "Turn On"
        on_press: flash.on()
    Button:
        text: "Turn Off"
        on_press: flash.off()
    Button:
        text: "Release"
        on_press: flash.release()
```

Figure 7-3. *The result of reading a text file in Python using Java classes*

After understanding the basic concept behind the Pyjnius library, we can use it to access features in Android.

Playing Audio in Android Using Pyjnius

We can start by discussing the Java code used to play audio in Android and think about how to write it in Python using Pyjnius. The code is shown in Listing 7-8, in which there is a single class used named MediaPlayer in the android.media package. After instantiating that class and returning an instance into the mediaPlayer variable, we can call the required methods to load and play the audio file.

The setDataSource() method accepts the filename of the audio file to be played. The prepare() method prepares the media player. Finally, the start() method starts playing the audio file. In order to catch the exceptions thrown when there is a problem playing the file, the call to such methods is bounded by a try-catch block. If there is an exception, a message is printed to the console using the System class.

Listing 7-8. Java Code to Play an Audio File in Android

```java
import android.media.MediaPlayer;

MediaPlayer mediaPlayer = new MediaPlayer();

try {
    fileName = "bg_music_piano.wav";
    mediaPlayer.setDataSource(fileName);
    mediaPlayer.prepare();
    mediaPlayer.start();
} catch {
    System.out.println("Error playing the audio file.");
}
```

After completing the Java code, we next need to discuss the process for reflecting this code in Python.

This class can be reflected in Python using the autoclass() function discussed before. If the object created from that reflected class is named mediaPlayer, the methods will be called exactly as was done in the Java code.

Regarding handling exceptions in Python, there are two points worth mentioning. The first one is that blocks in Python are defined using indentation. The second one is that exceptions in Java are handled inside the catch block. In Python, the block name is except. It is important to do such mappings between Python and Java. The Python code for playing an audio file using Pyjnius is shown in Listing 7-9.

Listing 7-9. Playing an Audio File in Python by Reflecting Java Classes Using Pyjnius

```python
import jnius

MediaPlayer = jnius.autoclass("android.media.MediaPlayer")

mp = MediaPlayer()

try:
    fileName = "bg_music_piano.wav"
    mp.setDataSource(fileName)
    mp.prepare()
    mp.start()
except:
    print("Error Playing the Audio File")
```

Based on the code in Listing 7-9, we can create an application for playing, pausing, and stopping audio. The KV file of the application is shown in Listing 7-10. There is a BoxLayout with five widgets. The first two widgets are Labels and the last three are Buttons.

The first Label is updated according to the progress of playing the audio file. This is done by drawing a rectangle inside canvas.before using the Rectangle vertex instruction, which is colored red according to the Color instruction. The default size of this Label widget is 0.0, meaning it is hidden when the application starts. The size changes according to the playing progress.

The second Label widget displays information while playing the file indicating the duration of the file in milliseconds, the position in milliseconds in which the file is playing, and the percentage of progress from 0.0% to 100.00%.

Listing 7-10. KV File for Playing an Audio File in Android

```
BoxLayout:
    orientation: "vertical"
    Label:
        id: audio_pos
        size_hint_x: 0.0
        size: (0.0, 0.0)
        canvas.before:
            Color:
                rgb: (1, 0, 0)
            Rectangle:
                pos: self.pos
                size: self.size
    Label:
        id: audio_pos_info
        text: "Audio Position Info"
    Button:
        text: "Play"
        on_release: app.start_audio()
    Button:
        text: "Pause"
        on_release: app.pause_audio()
    Button:
        text: "Stop"
        on_release: app.stop_audio()
```

The three buttons are responsible for starting, pausing, and stopping the audio. The callback functions start_audio(), pause_audio(), and stop_audio() are called when the on_release event is fired.

Figure 7-4 shows the window of the application after it runs on the Android device.

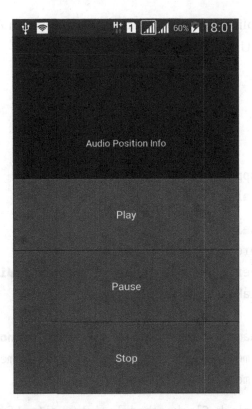

Figure 7-4. *An Android application for playing an audio file*

The Python file is shown in Listing 7-11. It implements the callback functions associated with the three Button widgets. The start_audio() function does the work discussed previously, from reflecting the MediaPlayer class until playing the audio file using the start() method. Notice that the instance of the MediaPlayer is set as a property inside the current object referenced using the self keyword in order to access it outside the function it is defined into. It can be accessed by self.mediaPlayer.

A class variable named prepare_audio is created to determine whether the audio file is previously loaded. It is initialized to False to mean that the file is not loaded. If its value is False, the start_audio() function loads the file and starts it. After being started, its value is set to True.

If the audio file is just paused, we do not have to reload the file again. Thus, the file will be started by calling the start() method inside the else part of the if statement. This resumes the audio file from the previous position before being paused. If the file is prepared after being paused, the file will play from the beginning.

Note that if there is a problem playing the audio file, the except block changes the text of the second Label to indicate an error.

Listing 7-11. Kivy Application for Playing an Audio File in Android Based on Reflected Java Classes Using Pyjnius

```
import kivy.app
import jnius
import os
import kivy.clock

class AudioApp(kivy.app.App):
    prepare_audio = False

    def start_audio(self):
        if AudioApp.prepare_audio == False:
            MediaPlayer = jnius.autoclass("android.media.MediaPlayer")
            self.mediaPlayer = MediaPlayer()
            try:
                fileName = os.getcwd()+"/bg_music_piano.wav"
                self.mediaPlayer.setDataSource(fileName)
                self.mediaPlayer.prepare()

                kivy.clock.Clock.schedule_interval(self.update_position, 0.1)

                self.mediaPlayer.start()
                AudioApp.prepare_audio = True
            except:
                self.current_pos.text = "Error Playing the Audio File"
                print("Error Playing the Audio File")
        else:
                self.mediaPlayer.start()

    def pause_audio(self):
        if AudioApp.prepare_audio == True:
            self.mediaPlayer.pause()

    def stop_audio(self):
        if AudioApp.prepare_audio == True:
            self.mediaPlayer.stop()
            AudioApp.prepare_audio = False
```

```
def update_position(self, *args):
    audioDuration = self.mediaPlayer.getDuration()
    currentPosition = self.mediaPlayer.getCurrentPosition()
    pos_percent = float(currentPosition)/float(audioDuration)

    self.root.ids['audio_pos'].size_hint_x = pos_percent

    self.root.ids['audio_pos_info'].text = "Duration: "
    +str(audioDuration) + "\nPosition: " + str(currentPosition)+
    "\nPercent (%): "+str(round(pos_percent*100, 2))

app = AudioApp()
app.run()
```

Using the schedule_interval() function inside the kivy.clock.Clock class, a callback function named update_position() is executed every 0.1 seconds. This updates the two Label widgets inside the KV file according to the current millisecond being played from the audio file.

First, the audio file duration is returned using the getDuration() method inside the audioDuration variable. The current position, in milliseconds, is returned using the getCurrentPosition() method inside the currentPosition variable. The progress percentage is calculated by dividing the audioDuration by currentPosition. Because the values inside these variables are integers and the expected result is between 0.0 and 1.0, which is float, we have to change their data type from integer to float using the float() function. The file duration, current position, and current percentage are displayed on the second Label widget inside the KV file.

Note that displaying the progress of playing the audio file is similar to updating the life percentage of the character of the CoinTex game. After discussing the function that handles the on_release event of the first button, we can discuss the functions for the remaining two buttons.

The second Button widget pauses the played audio file using the pause() method. According to its callback function, pause_audio(), an if statement ensures that the file is playing before executing this method because it must be called for an active audio file.

The last button stops the audio by calling the stop() method. Similar to the pause_audio() callback function, the stop_audio() function has an if statement that ensures that the audio file is playing before stopping it. After calling this method, the file can be played again only after preparing the file again. Thus, we must execute the code that prepares the audio file inside the if statement of the start_audio() function.

Figure 7-5 shows the result after playing the audio file and pausing it at millisecond 5943, which corresponds to a percentage equal to 19.44% of the complete file duration. The first label width is 19.44% of the screen width.

Figure 7-5. *Playing an audio file in Android by reflecting Java classes in Python*

This example easily used Java code within Python. Unfortunately, the process is not straightforward for even simple operations such as showing a toast message. In Java, a toast is displayed using just a single line, as shown:

```
Toast.makeText(this, "Hello Java", Toast.LENGTH_LONG).show();
```

The process requires more than reflecting the Toast class (android.widget.Toast) within Python. For example, we have to know that the text to be displayed is actually an instance of the CharSequence class. Thus, this class must be reflected in order to convert the text to its type to be suitable to the text argument. Moreover, toast is only created within the UI thread, not outside it. Thus, it can be displayed inside a Runnable instance. The developer has to do a lot of work than with the Java example.

Remember that the goal of using Pyjnius is to make life simpler for Kivy developers to build Android applications that use Java features in Python. If the process will get complicated, I do not recommend using it to write Java code. It can be used for simple tasks.

But if we cannot write Java code within Python, where do we write it? The answer is simply within a Java file.

Understanding the Android Project Built Using Buildozer

Remember that Python-for-Android is a bridge between Python (i.e., Kivy) and Android (i.e., Java). Using Buildozer, an Android Java project is automatically built from the Python Kivy project. After building the Android project, the Java project exists in the following path, assuming that the root directory of the Python project is NewApp and the package name specified inside the buildozer.spec file is kivyandroid.

`NewApp/.buildozer/android/platform/build/dists/kivyandroid`

If you are familiar with the structure of the Android Studio projects, you can navigate through the project to find all the necessary files and directories inside a regular project.

At the root directory of the project, the build.gradle and AndroidManifest.xml files exist. We will not discuss these files in detail but just get the idea that helps us understand how to manage the Android project from Android Studio.

The AndroidManifest.xml file is shown in Listing 7-12. It starts with the <manifest> element. Inside the header of this element, the package name and the application version are specified. It starts by specifying the application package name using the package attribute, which is set to com.test.kivyandroid. This is the result of concatenation between the package.domain and package.name properties inside the buildozer.spec file.

The android:versionCode property is an integer value representing the internal version number of the application to determine whether the version is newer than another. When uploading an application to Google Play, this integer helps it to know that there is a new version of the application to notify users to update it. But this integer is not displayed to the users. The value of the android:versionName attribute is a string, which is displayed to the users.

The android:installLocation allows us to specify where to install the application. It is set to auto to install the application on the external storage if the internal storage is full.

The <supports-screens> element does not set a restriction on the target screen and sets all of its attributes to True in order to support a wide range of devices.

The <uses-sdk> element specifies the minimum and target SDK using the android:minSdkVersion and android:targetSdkVersion attributes. Try to target the highest possible SDK versions, but set the target SDK to at least 26, as Google Play is no longer accepting applications with target SDKs less than 26.

The <uses-permission> element sets the required permissions by the application. Remember from Chapter 1 that there is a file called AndroidManifest.tmpl.xml inside the templates folder, located inside the Android project root directory, that loops through the android.permissions property in the buildozer.spec file in order to create a <uses-permission> element for each requested permission. The WRITE_EXTERNAL_STORAGE element is always requested by the application.

Listing 7-12. The AndroidManifest.xml File Inside the Kivy Android Studio Project

```
<?xml version="1.0" encoding="utf-8"?>
<manifest xmlns:android="http://schemas.android.com/apk/res/android"
     package="com.test.kivyandroid"
     android:versionCode="3"
     android:versionName="0.3"
     android:installLocation="auto">

  <supports-screens
          android:smallScreens="true"
          android:normalScreens="true"
          android:largeScreens="true"
          android:anyDensity="true"
          android:xlargeScreens="true" />

  <uses-sdk android:minSdkVersion="18" android:targetSdkVersion= "26" />

  <!-- OpenGL ES 2.0 -->
  <uses-feature android:glEsVersion="0x00020000" />
```

```xml
<!-- Allow writing to external storage -->
<uses-permission android:name="android.permission.WRITE_EXTERNAL_
STORAGE"/>

<uses-permission android:name="android.permission.CAMERA" />
<uses-permission android:name="android.permission.INTERNET" />

<application android:label="@string/app_name"
             android:icon="@drawable/icon"
             android:allowBackup="true"
             android:theme="@android:style/Theme.NoTitleBar.Fullscreen"
             android:hardwareAccelerated="true" >

    <meta-data android:name="wakelock" android:value="0"/>

    <activity android:name="org.kivy.android.PythonActivity"
              android:label="@string/app_name"
              android:configChanges="keyboardHidden|orientation|screen
              Size"
              android:screenOrientation="portrait" >

        <intent-filter>
            <action android:name="android.intent.action.MAIN" />
            <category android:name="android.intent.category.LAUNCHER" />
        </intent-filter>
    </activity>

</application>

</manifest>
```

When building an Android project, there will be an application that holds multiple activities (which are the Java classes). The AndroidManifest.xml file is structured to reflect that. It has an element named <application>, which declares the entire application. Using the <activity> element, we can declare each individual activity. The <application> element holds one or more of the <activity> elements as children.

Starting with the <application> element, it has a number of attributes in its header to define the application properties. For example, the application name is specified using the android:label attribute. This is set to a string that appears to the user. You can set it as a raw string, but it is better to set it as a string resource to reference it in other parts of the application, as we will see later.

In the Kivy built project, the string specifying the application name is added as a string resource with ID app_name. The strings.xml file inside the path shown here holds the string resources used by the application. This file has a template named strings.tmpl.xml inside the templates directory.

NewApp/.buildozer/android/platform/build/dists/kivyandroid/src/main/res/
values

This file is shown in Listing 7-13. There is an element named app_name, which holds the name of the application (CoinTex). In order to access the value inside the string resource named app_name within the AndroidManifest.xml file, we first need to refer to the strings.xml resource file using @string and then specify the ID of the string resource using @string/app_name.

The strings.xml file also has a number of string resources, such as presplash_color, which is the screen background shown while the application is starting.

Listing 7-13. The strings.xml File Inside the Kivy Android Studio Project

```xml
<?xml version="1.0" encoding="utf-8"?>
<resources>
    <string name="app_name">CoinTex</string>
    <string name="private_version">1544693478.42</string>
    <string name="presplash_color">#000000</string>
    <string name="urlScheme">kivy</string>
</resources>
```

The <application> element has another attribute named android:icon to set the application icon. It is set as a drawable resource with ID icon. The drawable resource file is located in the following path. Note that the ID icon refers to an image (bitmap file) named icon with an extension, which could be .png, .jpg, or .gif.

NewApp/.buildozer/android/platform/build/dists/kivyandroid/src/main/res/
drawable

If you set the fullscreen property of the buildozer.spec file to 1, the android:theme attribute will apply a style that hides the notification bar.

The `<application>` element has another attribute named `android:hardware Accelerated`, which is set to `True` to smoothly render the graphics displayed on the screen.

The `<activity>` element declares the activities within the application. There is just a single activity named `org.kivy.android.PythonActivity`, which is the main activity. The Java class for this activity is located in the following path. By locating the Java classes of the application, we can add whatever Java code needs to be executed. We will see this later.

NewApp/.buildozer/android/platform/build/dists/kivyandroid/src/main/java/ org/kivy/android

Inside the `<activity>` element header, the `android:label` attribute is set to the value inside the string resource with ID `app_name`. This is the same resource used to set the application name inside the `<application>` element that is reused. This is why it is important to use string resources.

When some events happen, such as hiding the keyboard while an activity is running, the activity is restarted by default. The `android:configChanges` attribute determines the set of configuration changes to be handled by the activity without being restarted. In our project, three values are assigned to this attribute, which are `keyboardHidden` in addition to `orientation` and `screenSize`, for handling screen orientation changes from landscape to portrait and vice versa.

The `android:screenOrientation` attribute sets the orientation of the device. It reflects the value stored in the `orientation` property in the `buildozer.spec` file.

The `<activity>` element has a child element called `<intent-filter>`, which declares the intent of the activity. This is what the activity can do. Other applications can call your activity for using its capacities declared within that element. The `<intent-filter>` element must have at least one `<action >` element in order to allow it to accept intents. In our activity, its action is shown the name `android.intent.action.MAIN`. This means the parent activity is the entry point of the application (i.e., the activity opens after opening the application). In order to list the application in the application launcher of the device, the `<category>` element is added with a name equal to `android.intent. category.LAUNCHER`.

This is a quick overview of the `AndroidManifest.xml` file. Note that the values inside that file may be overridden by the `build.gradle` file. The interesting part inside that file is the one declaring the minimum SDK and target SDK, as shown in Listing 7-14. It is easy to understand that the minimum SDK version is 18, the target SDK is 26, the version code is integer 3, and the version name is the string 0.3. Note that these values are identical to those defined inside the `AndroidiManifest.xml` file.

Remember that there is a template file called `build.tmpl.gradle` that accepts these values from the `buildozer.spec` file and produces the `build.gradle` file. It is also located inside the `templates` directory within the `AndroidManifest.tmpl.xml` and `strings.tmpl.xml` files.

Listing 7-14. Specifying the Minimum SDK and Target SDK Versions Inside the build.gradle File

```
android {
    compileSdkVersion 18
    buildToolsVersion '28.0.3'
    defaultConfig {
        minSdkVersion 18
        targetSdkVersion 26
        versionCode 3
        versionName '0.3'
    }
```

When manipulating the Android project within Android Studio, Android Studio will search for the values of the `buildToolsVersion` and `compileSdkVersion` fields specified in the `build.gradle` file. If they aren't found, the project will not be built. You can change these versions to whatever works with your system.

Project Main Activity

According to the previous section, the main activity class is named `PythonActivity` and it is located inside the `org.kivy.android` package. The class is not an activity due to extending the `Activity` class (`android.app.Activity`) but due to extending the `SDLActivity` class, which in turn extends the `Activity` class according to the class header shown here.

```
public class PythonActivity extends SDLActivity{}
```

The SDLActivity class exists in the org.libsdl.app package. The file is located in the path shown next. Note that we are using SDL as the backend of Kivy and this is why the class has SDL in its name. If you are using PyGame, there might be some changes.

NewApp/.buildozer/android/platform/build/dists/kivyandroid/src/main/java/
org/libsdl/app

Because the entry point for any Android application is the onCreate() method, we can discuss it. A few lines from the top of the PythonActivity class in addition to the method are shown in Listing 7-15. The first two lines import the classes named SDLActivity and ResourceManager.

We used to find a class named R.java in the Android projects to manage the resources by maintaining their IDs for example. For Kivy projects, this class is replaced by a class called ResourceManager, which is in the org.renpy.android package. This is why an instance is created from this class at the beginning of the onCreate() method.

A public static class variable named mActivity is created and initialized to null. Inside the onCreate() method, this variable is set to refer to the current activity as it is assigned the current object, this, referring to the activity. Note that we can use this variable to access the main activity using Pyjnius inside a Python script. This is done by reflecting the PythonActivity class and then accessing its property, mActivity.

Listing 7-15. The onCreate() Method Inside the PythonActivity Class

```java
import org.libsdl.app.SDLActivity;
import org.renpy.android.ResourceManager;
import android.os.Bundle;
...
public class PythonActivity extends SDLActivity {
    private static final String TAG = "PythonActivity";
    public static PythonActivity mActivity = null;
    ...

    @Override
    protected void onCreate(Bundle savedInstanceState) {
        Log.v(TAG, "My oncreate running");
        resourceManager = new ResourceManager(this);
```

```
        Log.v(TAG, "About to do super onCreate");
        super.onCreate(savedInstanceState);
        Log.v(TAG, "Did super onCreate");

        this.mActivity = this;
        this.showLoadingScreen();

        new UnpackFilesTask().execute(getAppRoot());

        Toast.makeText(this, "Working on the Kivy Project in Android
        Studio", Toast.LENGTH_LONG).show();
    }
...
}
```

A method named showLoadingScreen() inside the activity is called. This method just loads the presplash image inside an ImageView. Remember that the name of this image is specified inside the presplash.filename property of the buildozer.spec file. That method is shown in Listing 7-16.

The method works as follows. If the ImageView is not created inside the mImageView variable, it creates it by loading the presplash image resource. The resource identifier is returned in the presplashId variable using the getIdentifier() method inside the resourceManager instance. This method accepts the resource name and its kind and returns an integer representing the ID. This is why the presplashId variable type is integer. Note that in regular Android projects, the findViewById() method returns the IDs.

The presplash image is located in this path NewApp/src/main/res/drawable/ presplash.jpg. Because the resource is an image, its kind is drawable and thus it's added to the drawable directory. This directory is located under the res directory to indicate that it holds resources. The name of a resource is presplash, which is the resource filename without the extension.

After returning its ID, the raw image file is opened and returned to the is variable, which is an instance of the InputStream class. Then the raw data is decoded as an image using the decodeStream() method of the BitmapFactory class. The data is returned to a variable named bitmap.

After that, an instance of the ImageView class is returned to the mImageView variable and the image displayed on it is set using the setUmageBitmap() method. It accepts the bitmap variable we previously created.

Remember that the background color of the loading screen is saved as a string resource inside the strings.xml file with an ID of presplash_color. To return the value inside a string resource, the getString() method inside the ResourceManager class is used. Using the setBackgroundColor() method of the ImageView class, the background color of the mImageView instance is changed. Some parameters are specified for the ImageView in order to fill the parent size.

Listing 7-16. The showLoadingScreen() Method

```
protected void showLoadingScreen() {
    if (mImageView == null) {
        int presplashId = this.resourceManager.getIdentifier("presplash",
        "drawable");
        InputStream is = this.getResources().openRawResource(presplashId);
        Bitmap bitmap = null;
        try {
            bitmap = BitmapFactory.decodeStream(is);
        } finally {
            try {
                is.close();
            } catch (IOException e) {};
        }

        mImageView = new ImageView(this);
        mImageView.setImageBitmap(bitmap);

        String backgroundColor = resourceManager.getString("presplash_
        color");
        if (backgroundColor != null) {
            try {
                mImageView.setBackgroundColor(Color.parseColor(background
                Color));
            } catch (IllegalArgumentException e) {}
        }
        mImageView.setLayoutParams(new ViewGroup.LayoutParams(
                ViewGroup.LayoutParams.FILL_PARENT,
                ViewGroup.LayoutParams.FILL_PARENT));
```

```
        mImageView.setScaleType(ImageView.ScaleType.FIT_CENTER);
    }

    if (mLayout == null) {
        setContentView(mImageView);
    } else if (PythonActivity.mImageView.getParent() == null){
        mLayout.addView(mImageView);
    }
}
```

There is a variable named mLayout that refers to the activity layout holding all views. This variable is defined inside the SDLActivity class as protected and static, according to the next line:

protected static ViewGroup *mLayout*;

In order to display the presplash image view, the image view must be added to the activity layout. In order to add a view to the layout, the addView() method is used. Using the command mLayout.addView(mImageView), the presplash image view is added to the activity layout referred to it by the mLayout variable.

It is very important to note the following:

- There is no layout XML resource file for the application. Thus, the layouts are created dynamically within the Java code.

- There is just a single layout for the activity defined in the static mLayout variable of the SDLActivity class. As a result, the views added to the layout are visible until we remove them. For example, the presplash image view must be removed from the layout before the application starts (i.e., after loading ends).

After the screen finishes loading, the presplash image view is removed according to the method named removeLoadingScreen() inside the PythonActivity class. It will be discussed later in this chapter.

Note that if there is no layout created previously, the image itself is displayed on the screen as the main view using the setContentView() function.

The purpose of displaying the presplash image on the loading screen rather than starting the application directly is that files must be loaded before the application starts. Inside the project root directory, there is a folder named python-install that includes all the Python files needed to build the application.

After displaying the loading screen, the onCreate() method starts a background thread by creating an instance from the UnpackFilesTask class that extends AsyncTask. Note that a toast message is shown after running this thread while the loading screen is active.

The class header and some chunks of code from this class are shown in Listing 7-17. Note that this class is nested inside the PythonActivity class and thus no import statement is required.

The UnpackFilesTask class implements the doInBackground() callback method inside the AsyncTask class, which starts as soon as the instance is created. Inside it, a method called unpackData() inside the UnpackFilesTask class is called. It loads the project files.

Listing 7-17. The UnpackFilesTask Class

```
import android.os.AsyncTask;
    ...
private class UnpackFilesTask extends AsyncTask<String, Void, String> {
    @Override
    protected String doInBackground(String... params) {
        ...
        unpackData("private", app_root_file);
        return null;
    }

    @Override
    protected void onPostExecute(String result) {
        ...
        mActivity.finishLoad();

        mActivity.showLoadingScreen();
    }
...
}
```

After the files are loaded, a method named keepActive() is called by SDL. It is shown in Listing 7-18. This method is defined in the SDLActvity class and implemented inside the PythonActivity class.

Listing 7-18. Implementation of the keepActive() Method

```
@Override
public void keepActive() {
  if (this.mLoadingCount > 0) {
    this.mLoadingCount -= 1;
    if (this.mLoadingCount == 0) {
      this.removeLoadingScreen();
    }
  }
}
```

This method calls another method named removeLoadingScreen(), which removes the presplash image view from the activity layout. Its implementation is shown in Listing 7-19. It creates a thread that runs on the application UI. Within this thread, the parent of the mImageView variable is returned using the getParent() method, which refers to the activity layout. Using the removeView() method, mImageView is removed from the layout. In order to remove that value inside this variable, its value is returned to null.

Listing 7-19. Implementation of the removeLoadingScreen() Method

```
public void removeLoadingScreen() {
  runOnUiThread(new Runnable() {
    public void run() {
      if (PythonActivity.mImageView != null &&
          PythonActivity.mImageView.getParent() != null) {
        ((ViewGroup)PythonActivity.mImageView.getParent()).removeView(
        PythonActivity.mImageView);
        PythonActivity.mImageView = null;
      }
    }
  });
}
```

After executing the removeLoadingScreen() method, the activity layout will be empty. This means we are ready to fill the layout by the new UI elements, as defined inside in the Python application. For such purposes, the UnpackFilesTask class also implements the onPostExecute() callback method, which gets executed when the class thread finishes execution. Such method is executed on the UI thread and thus able to manage the UI. Inside the method, another method named finishLoad() is executed. This method is defined in the SDLActivity class. This class header and some lines of the finishLoad() method are both shown in Listing 7-20.

Listing 7-20. Setting the Application Layout Inside the finishLoad() Method Within the SDLActivity Class

```
import android.widget.AbsoluteLayout;
import android.view.*;
import android.app.*;
    ...
public class SDLActivity extends Activity {
    ...
    protected static ViewGroup mLayout;
    ...

    protected void finishLoad() {
        ...
        mLayout = new AbsoluteLayout(this);
        mLayout.addView(mSurface);

        setContentView(mLayout);
    }
...
}
```

Note that the goal of finishing the loading screen is not just starting the application. Another purpose is to make the UI layout managed by the SDLActivity, as it is the one responsible for preparing the application UI. The finishLoad() method prepares a layout inside the mLayout instance. We will use this method later for adding views to the layout of the main activity (PythonActivity).

Displaying Toast Messages from Python

Inside the PythonActivity class, there is a method called toastError(), as shown in Listing 7-21. It shows a toast message according to an input string argument. It runs a thread on the UI using runOnUIThread(). This method is called using the PythonActivity instance, which is saved into the mActivity variable.

Listing 7-21. Implementation of the toastError() Method to Display Toast Messages

```
public void toastError(final String msg) {

    final Activity thisActivity = this;

    runOnUiThread(new Runnable () {
        public void run() {
            Toast.makeText(thisActivity, msg, Toast.LENGTH_LONG).show();
        }
    });

    // Wait to show the error.
    synchronized (this) {
        try {
            this.wait(1000);
        } catch (InterruptedException e) {
        }
    }
}
```

We can benefit from this method to display toast messages inside the Python code. The only thing to do is to reflect the PythonActivity class and access its mActivity property, which will call the toastError() method. The KV file of the application is shown in Listing 7-22. The root BoxLayout has a button that calls a method called show_toast() when released. Because the toast color is dark, a label is added in white in order to make it easier to see the toast message.

Listing 7-22. KV File for Displaying a Toast Method Using Python

```
BoxLayout:
    orientation: "vertical"
    Button:
        text: "Show Toast"
        on_release: app.show_toast()
    Label:
        canvas.before:
            Color:
                rgb: (1, 1, 1)
            Rectangle:
                pos: self.pos
                size: self.size
```

The Python file that shows the toast is shown in Listing 7-23. The PythonActivity is reflected only once using jnius.autoclass. The mActivity property is returned from the class. Inside the show_toast() method, the toastError() method is called using the mActivity variable.

Listing 7-23. Displaying a Toast Message Within Python by Reflecting the PythonActivity Class

```
import kivy.app
import jnius
import kivy.uix.button

PythonActivity = jnius.autoclass("org.kivy.android.PythonActivity")
mActivity = PythonActivity.mActivity

class ToastApp(kivy.app.App):

    def show_toast(self):
        mActivity.toastError("Test Toast :)")

app = ToastApp()
app.run()
```

After releasing the button, the toast is displayed according to Figure 7-6.

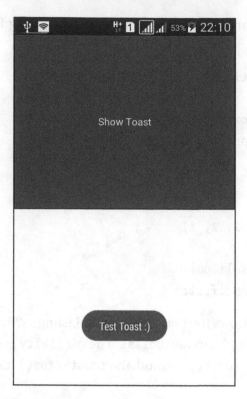

Figure 7-6. *A toast message displayed using Python*

Opening Kivy Android Projects in Android Studio

From the previous discussion, we at least have a basic understanding of how the Android project created by Buildozer works. The next step is to edit this project in Android Studio.

Up to this section, Linux has been used to develop the projects. This is because Buildozer is only supported in Linux. After generating the Android project, we can use the operating system of our choice for manipulating it in Android Studio. In this section, Windows is used to import and edit the project. Before starting, make sure that Android Studio is working correctly and you have the versions of build tools and SDK listed in the `gradle.build` file.

Before editing the project, we need to import it. Simply choose the Open item from the File menu. A window appears, as shown in Figure 7-7, for navigating to the project's path. The Android Studio icon will appear next to the project name, which means it is a project that Android Studio understands. Click OK to open the project.

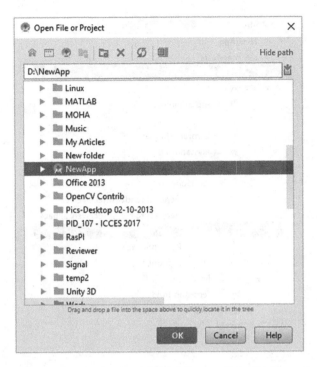

Figure 7-7. *Importing the Kivy Android Studio project built using Buildozer within Android Studio*

After the project opens, the project files will appear on the left side of the window. The structure of the files in the Android view is shown in Figure 7-8. This view helps you view the files in a simple way regardless of their actual locations inside the project. For example, the unnecessary files are not shown and any related files are grouped together.

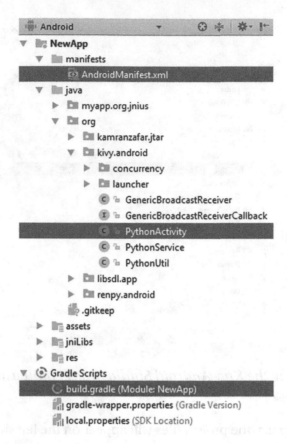

Figure 7-8. *The structure of the imported project within Android Studio*

Inside the `manifests` group, all manifests used within the project are listed. This project has a single manifest file called `AndroidManifst.xml`. The Java main activity class (`PythonActivity`) is located in the `java` group. We can easily deduce that this class is inside the `org.kivy.android` package. The `build.gradle` file is within a group named `Gradle Scripts`. The `res` group contains all the resources, such as the `strings.xml` file. You can view the project files as stored on the disk by selecting the `Project` view.

After importing the project, we can run the project either to an emulator or an USB connected device according to Figure 7-9.

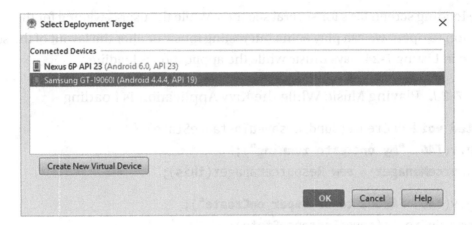

Figure 7-9. *Selecting a USB-connected device for running the Android application*

After we click OK, the project will be built, deployed, installed, and run on the selected device. Figure 7-10 shows the modified loading screen where the toast message is displayed. If everything works up to this point, then we did a good job.

Figure 7-10. *Modifying the Kivy presplash image and displaying a toast message while the application is loading*

The loading screen lasts for several seconds. While the user is waiting for the application to open, we can play some interesting music or alter the layout of the screen. The code in Listing 7-24 plays music while the application is loading.

Listing 7-24. Playing Music While the Kivy Application Is Loading

```
protected void onCreate(Bundle savedInstanceState) {
    Log.v(TAG, "My oncreate running");
    resourceManager = new ResourceManager(this);

    Log.v(TAG, "About to do super onCreate");
    super.onCreate(savedInstanceState);
    Log.v(TAG, "Did super onCreate");

    this.mActivity = this;
    this.showLoadingScreen();

    new UnpackFilesTask().execute(getAppRoot());

    Toast.makeText(this, "Working on the Kivy Project in Android Studio",
    Toast.LENGTH_LONG).show();

    int music_id = resourceManager.getIdentifier("music", "raw");
    MediaPlayer music = MediaPlayer.create(this, music_id);
    music.start();
}
```

The music file is added as a war to the Android project. This is by creating a directory within the res group and adding the music file inside it, according to Figure 7-11. Adding resources helps to retrieve them easily within the application. This is better than referencing resources using their path in the device.

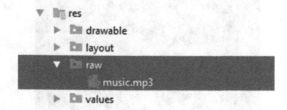

Figure 7-11. *Adding the music file as a raw resource inside the Android Studio project*

After adding the raw resource into the project, we must reference it within the Java code. Remember that resources in Kivy Android projects are manipulated using a class named ResourceManager inside the org.renpy.android package.

In the previous section, the ID of the drawable resource named presplash is returned using the getIdentifier() method inside that class. Similarly, the identifier of the raw resource is returned using this method. Just specify the suitable name and kind arguments.

Because the resource filename is music.mp3, the name argument is set to music. Its kind is set to raw because it exists in the raw folder. Because IDs of resources are integers, the ID is returned to the music_id variable of type integer.

After running the application, the music will be played as soon as the loading screen appears. It will automatically be stopped after the files are done loading.

Adding Views to the PythonActivity Layout

By default, the presplash image is the only view displayed on the activity layout while the screen is loading. We can alter this layout to add more views. Just remember that there is only one layout for the activity and thus we have remove the newly added views once they are not required.

The place where we can add more views while the screen is loading is the showLoadingScreen() method of the PythonActivity class. We can add a single TextView within the activity layout according to this modified method shown in Listing 7-25. The TextView is added after the presplash image view that's defined into the mImageView variable gets added to the layout. The code omits some parts and focuses on the part where the TextView is added to the layout.

An instance from the TextView class is returned into a static variable defined in the PythonActivity class named loadingTextView. The reason for not making this variable local to the method is that we need to access it later inside the removeLoadingScreen() method to remove it.

The text displayed in the text view is specified using the setText() method. Its text size is changed using the setTextSize() method. After preparing the text view, it is added to the layout using the addView() method.

Listing 7-25. Adding a TextView to the Loading Screen While the Kivy Application Is Loading

```
static TextView loadingTextView = null;
    ...
  protected void showLoadingScreen() {
    ...
  if (mLayout == null) {
      setContentView(mImageView);
  } else if (PythonActivity.mImageView.getParent() == null){
      mLayout.addView(mImageView);

      // Adding Custom Views to the Layout
      loadingTextView = new TextView(this);
      loadingTextView.setText("Kivy application is loading. Please wait ...");
      loadingTextView.setTextSize(30);

      mLayout.addView(loadingTextView);
  }
}
```

After we run the application, the text view will appear as shown in Figure 7-12.

Figure 7-12. *Adding an Android TextView to the loading screen of the Kivy application*

Assuming that the text view is not removed from the layout, it will remain visible in the screen after the application starts, as shown in Figure 7-13.

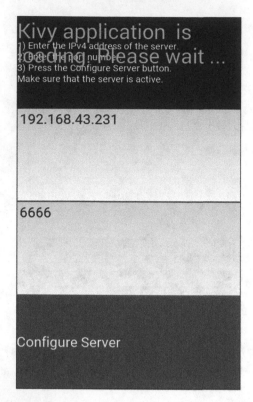

Figure 7-13. *The TextView added to the application layout while it is loading remains visible after the application starts*

Remember that this happens because there is only one layout for the application that is used in both the loading screen and after the applications starts. Thus, it is very important to remove that view inside the removeLoadingScreen() method, as shown in Listing 7-26. Similar to removing presplash image view, the parent of the text view is returned. The parent calls the removeView() method to remove it.

Listing 7-26. Removing the TextView Inside the removeLoadingScreen() Method

```
public void removeLoadingScreen() {
  runOnUiThread(new Runnable() {
    public void run() {
      if (PythonActivity.mImageView != null &&
          PythonActivity.mImageView.getParent() != null) {
        ((ViewGroup)PythonActivity.mImageView.getParent()).removeView(
        PythonActivity.mImageView);
```

```
        PythonActivity.mImageView = null;
        ((ViewGroup)PythonActivity.loadingTextView.getParent()).removeView(
        PythonActivity.loadingTextView);
    }
  }
});
}
```

In summary, a view is added to the activity layout that remains as long as the screen is loading. Once the loading process is over, the view is removed. If the view is not removed from the parent, it will still visible after the application UI defined in Python is loaded.

Assume we need to add a view to the activity layout after the loading step ends. Where in the Java code do we add this view? Let's discuss this matter in the next section.

SDLSurface

In order to determine the suitable place within the project to add views after the files are loaded but before the application layout is visible, it is important to discuss the SDLSurface class.

Inside the SDLActivity Java file, there is a class named SDLSurface that extends the android.view.SurfaceView Android class, according to the SDLSurface class header shown here.

class SDLSurface **extends** SurfaceView **implements** SurfaceHolder.Callback, View.OnKeyListener, View.OnTouchListener, SensorEventListener

The SurfaceView is an Android component in which we can draw on the application UI. The UI widgets specified in the Kivy project KV file are drawn within this SurfaceView.

SurfaceView implements an interface named SurfaceHolder.Callback, which provides a number of callback methods to help receive information about the surface. For example, when the surface is created, the callback method surfaceCreated() is called. The surfaceDestroyed() and surfaceChanged() callback methods are called when the surface is destroyed and changed, respectively. An example of changes that occur is creating an instance of this class and drawing something on the surface.

To handle hardware key presses, SurfaceView implements the second interface named View.OnKeyListener. It has a single callback method named onKey()

that is called when a hardware key is pressed. To handle touch events, the View. OnTouchListener is implemented. It has a single callback method named onTouch().

The final interface implemented by the SurfaceView class is SensorEventListener. It listens to the changes in the sensor data. It has two callback methods— onAccuracyChanged() and onSensorChanged(). Both are implemented in the class but the onAccuracyChanged() is empty. The onSensorChanged() is used to monitor the device orientation.

As a summary of all methods defined inside the SDLSurface class, its header and the methods signatures are shown in Listing 7-27. It helps to have a basic idea about the components of the SDLSurface class. You can check out the implementation of all of these methods to get more information about how the class works.

Listing 7-27. Methods Inside the SDLSurface Class

```
class SDLSurface extends SurfaceView implements SurfaceHolder.Callback,
    View.OnKeyListener, View.OnTouchListener, SensorEventListener  {
    ...
    public SDLSurface(Context context) {
        ...
    }

    public void handleResume() {
        ...
    }

    public Surface getNativeSurface() {
        ...
    }

    @Override
    public void surfaceCreated(SurfaceHolder holder) {
        ...
    }

    @Override
    public void surfaceDestroyed(SurfaceHolder holder) {
        ...
    }
```

```java
@Override
public void surfaceChanged(SurfaceHolder holder, int format, int width,
int height) {
    ...
        final Thread sdlThread = new Thread(new SDLMain(), "SDLThread");
    ...
        sdlThread.start();
    ...
}

// unused
@Override
public void onDraw(Canvas canvas) {}

// Key events
@Override
public boolean onKey(View  v, int keyCode, KeyEvent event) {
    ...
}

// Touch events
@Override
public boolean onTouch(View v, MotionEvent event) {
    ...
}

// Sensor events
public void enableSensor(int sensortype, boolean enabled) {
    ...
}

@Override
public void onAccuracyChanged(Sensor sensor, int accuracy) {
    // TODO
}
```

```
    @Override
    public void onSensorChanged(SensorEvent event) {
        ...
    }
}
```

There are two lines that are worth mentioning inside the implemented
surfaceChanged() callback method of the SurfaceHolder.Callback interface. These
lines create and start an instance of the Thread class into the sdlThread variable, which
refers to the main SDL thread. There are two arguments specified inside the Thread class
constructor, which are the target Runnable object and the thread name.

The target is an instance of the SDLMain class that extends the Runnable interface. Note
that the SDLMain class exists inside the SDLActivity Java file. It is shown in Listing 7-28.

Listing 7-28. Implementation of the SDLMain Class

```
class SDLMain implements Runnable {
    @Override
    public void run() {
        SDLActivity.nativeInit(SDLActivity.mSingleton.getArguments());
    }
}
```

This class implements the run() method, which calls a static function named
nativeInit(). Its header is defined inside the SDLActivity class header, as shown here.

```
public static native int nativeInit(Object arguments);
```

This is not a Java method but a C function. This method is the entry point for the SDL
library. This method and the other C methods are found in the source file of the SDL
named libSDL2.so, located in the /NewApp/libs/rmeabi-v7a directory.

By reaching this step, the C files are ready to be executed and the application will
start and get ready for user interaction. We can return to our question, which is where in
the Java project do we add views that appear in the application activity after the screen
loading completes. Let's follow the chain of execution to find the suitable position.

It is now clear that the nativeInit() function inside the SDLMain class is the entry
point for executing the SDL library and rendering the application UI. Thus, we have
to add our views before calling this function. Back in the chain, an instance of the
SDLMain class is created within the surfaceChanged() method inside the SDLSurface

class. Remember that the surfaceChanged() method is a callback method inside the SurfaceHolder.Callback interface that listens to changes in the SurfaceView.

Because we want to add views to the application layout once it is started, we need to find the place in which an instance of the SurfaceView class is created. This is the first change to occur for such class. Remember that there is a class named SDLSurface that extends the SurfaceView class. An instance of the SDLSurface is created inside the finishLoad() method inside the SDLActivity class. The surface is stored in a variable named mSurface, which is declared as static and protected in the class header according to this line:

protected static SDLSurface *mSurface*;

The part of the finishLoad() method that creates such an instance is shown in Listing 7-29. After it's created, it is added as a child view inside the application layout stored in the mLayout variable. Finally, the layout is set as the activity layout using the setContentView() method. If we need to add a view to the activity layout, it can be added before calling the setContentView() method.

Listing 7-29. Instantiating the SDLSurface Class Within the finishLoad() Method

```
protected void finishLoad() {
    ...
    mSurface = new SDLSurface(getApplication());
    ...
    mLayout = new AbsoluteLayout(this);
    mLayout.addView(mSurface);

    setContentView(mLayout);
}
```

Remember that the finishLoad() method is called within the onPostExecute() callback method of the UnpackFilesTask class. That is after the application loads its files. We can assume that the finishLoad() method of the SDLActivity class is the starting point where the activity layout is prepared. It is a good idea to add views to the activity layout that is visible only after the application loading step ends.

At the end of the finishLoad() method, as shown in Listing 7-30, two views are created, which are TextView and Button. The text changes size using the setTextSize() method and the color is set to red using the setTextColor() method.

Listing 7-30. Adding a TextView and a Button to the Application Layout Inside the finishLoad() Method

```java
protected void finishLoad() {
    ...
    // Set up the surface
    mSurface = new SDLSurface(getApplication());

    if(Build.VERSION.SDK_INT >= 12) {
        mJoystickHandler = new SDLJoystickHandler_API12();
    }
    else {
        mJoystickHandler = new SDLJoystickHandler();
    }

    mLayout = new AbsoluteLayout(this);
    mLayout.addView(mSurface);

    // Adding Custom Views to the Layout
    TextView appTextView = new TextView(this);
    appTextView.setText("Loaded successfully.");
    appTextView.setTextColor(Color.parseColor("#ff0000"));
    appTextView.setTextSize(20);

    Button appButton = new Button(this);
    appButton.setText("Show Toast");
    appButton.setTextColor(Color.parseColor("#ff0000"));
    appButton.setTextSize(20);

    appButton.setOnClickListener(new View.OnClickListener() {
        @Override
        public void onClick(View view) {
            Toast.makeText(getApplicationContext(), "Toast on Java Button
            Click.", Toast.LENGTH_LONG).show();
        }
    });

    LinearLayout newLayout = new LinearLayout(this);
    newLayout.setOrientation(LinearLayout.HORIZONTAL);
```

```
newLayout.addView(appTextView);
newLayout.addView(appButton);

mLayout.addView(newLayout);

setContentView(mLayout);
}
```

The button click action is handled by specifying the click listener using
the setOnClickListener() method. This method accepts an instance of the
OnClickListener() interface after implementing its onClick() method. The code inside
this method is executed when the button is clicked. The button shows a toast message
when clicked.

The two new views are added as children to a parent LinearLayout with horizontal
orientation. Finally, the linear layout is added as a child to the activity layout.

Figure 7-14 shows the result after clicking the new button that was added in Java.

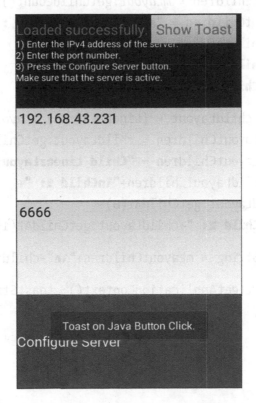

Figure 7-14. *Adding Android views to the Kivy application*

We can modify the toast message to print all child views inside the activity layout when the button is clicked. The code in Listing 7-31 modifies the onClickListener() callback method for that purpose. The number of child views inside the layout is returned using the getChildCount() method. To retrieve a specific view, the getChildAt() method accepts an index referring to the position of the view within the parent, where index 0 refers to the first child added to the parent.

Similarly, the child views defined inside the LinearLayout created in the previous example in addition to their number are shown.

Listing 7-31. Editing the onClickListener() Callback Method of the Button View to Print All Layout Child Views Inside a Toast Message

```
appButton.setOnClickListener(new View.OnClickListener() {
    @Override
    public void onClick(View view) {
        int numLayoutChildren = mLayout.getChildCount();
        String mLayoutChildren = "Num layout children :
        "+numLayoutChildren+
                "\nChild 1: "+mLayout.getChildAt(0)+
                ",\nChild 2: "+mLayout.getChildAt(1);

        LinearLayout childLayout = (LinearLayout) mLayout.getChildAt(1);
        int numChildLayoutChildren = childLayout.getChildCount();
        String childLayoutChildren = "Child LinearLayout children : "+
                numChildLayoutChildren+"\nChild 1: "+
                childLayout.getChildAt(0)+
                ",\nChild 2: "+childLayout.getChildAt(1);

        String toastString = mLayoutChildren+"\n"+childLayoutChildren;

        Toast.makeText(getApplicationContext(), toastString, Toast.LENGTH_
        LONG).show();
    }
});
```

Figure 7-15 shows the toast message after the button is clicked. It shows that the number of children in the layout is 2. The first one is an instance of the SDLSurface class and the second one is an instance of the LinearLayout class. The children inside the LinearLayout are also printed. The SDLSurface view contains all widgets defined in the Kivy application. In a later section, we discuss how events are handled for such widgets.

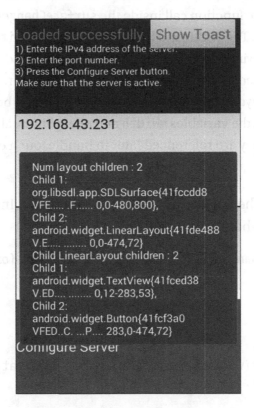

Figure 7-15. *A toast message for printing information about the child views inside the layout once the Button view is clicked*

In summary, in order to add views to the layout while the screen is loading, add the code inside the showLoadingScreen() method inside the PythonActivity class. The keepAlive() method is called by SDL when the loading completes, which calls a method named removeLoadingScreen() where views can be deleted from the layout after the loading. After the project files are loaded, the onPostExecute() callback method of the UnpackFilesTask class calls a method named finishLoad() inside the SDLActivity class. It is a good place to add views to the layout after the screen is loaded.

How Does SDL Detect the UI Widgets Defined in Kivy?

You might have noticed that the button added in the previous example hides part of the word server displayed on the Label widget defined in Kivy. Why does that happen?

When the surface is created, its size is set to the entire window according to the onNativeResize() native function call inside the surfaceChanged() callback method, as shown in Listing 7-32. The width and height of the SDLActivity surface are set to the width and height of the window. This does not leave room for adding new views inside Java. The views are stacked on the top of each other. The Z order of the latest views to be added into the activity layout appear on top of the views added before them.

It is important to set the variables mWidth and mHeight equal to the width and height of the surface. They are referenced later to handle touch events of the widgets added in Kivy.

Listing 7-32. Setting the Surface Size to the Window Size Inside the SurfaceChanged() Method

```
public void surfaceChanged(SurfaceHolder holder, int format, int width, int
height) {
    ...
    mWidth = width;
    mHeight = height;
    SDLActivity.onNativeResize(width, height, sdlFormat, mDisplay.
    getRefreshRate());
    ...
}
```

In the previous example, the new linear layout (newLayout) is added to the activity layout (mLayout) after that SDL surface (mSurface) is added. This is why the new views appear on top of the SDL surface.

The orientation of LinearLayout is set to horizontal. For horizontal linear layout, the default width and height are set to WRAP_CONTENT, which means the linear layout will not fill the entire activity layout size. It just covers the least space for holding its child views. This is why SDLSurface is still visible.

If the SDLSurface is added to the activity layout before the linear layout according to the finishLoad() method in Listing 7-33, then the TextView and Button inside the linear layout will be hidden by the SDL surface.

Listing 7-33. Adding the SDL Surface to the Activity Layout Before Adding the Custom LinearLayout

```
protected void finishLoad() {
    ...
    mLayout.addView(newLayout);
    mLayout.addView(mSurface);
    ...
}
```

One idea that comes to our mind to give each view its unique space is to change the SDL surface size to not fill the entire height of the window. The addView() method can accept another argument that specifies the layout parameters of a given view when being added to the layout. When adding the SDL surface to the layout inside the finishLoad() method, we can change its height to be 3/4 of the layout height, according to Listing 7-34.

The screen width and height are returned to the width and height variables. Because the activity layout fills its screen, we can use them interchangeably. They are required in order to position the SDL surface relative to the layout. The addView() method accepts the layout parameters as a second argument. Note that the AbsoluteLayout class is used because the activity layout is an instance of this class.

Listing 7-34. Setting the Height of the SDL Surface to be 3/4 of the Layout Height

```
protected void finishLoad() {
    ...
    // Return screen size to position the SDL surface relative to it.
    DisplayMetrics displayMetrics = new DisplayMetrics();
    getWindowManager().getDefaultDisplay().getMetrics(displayMetrics);
    int width = displayMetrics.widthPixels;
    int height = displayMetrics.heightPixels;
```

```
// Specifying the new height and Y position of the SDL surface.
mLayout.addView(mSurface, new AbsoluteLayout.LayoutParams(width,
height-height/4, 0, height/4));

...
}
```

The width of the surface is set equal to the screen width, but the height is set to
`height-height/4`. That is 3/4 of the layout height is for the SDL surface and 1/4 of the
activity layout height is empty and will hold the linear layout. The layout starts at the
top-left corner (0, 0). If this position is not changed, the SDL surface and the new linear
layout will appear on top of each other, as shown in Figure 7-16. The empty region
appears at the bottom of the layout.

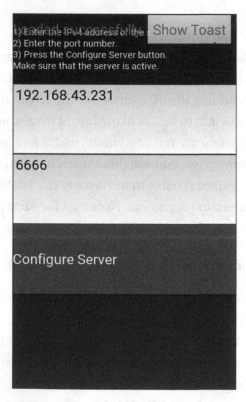

Figure 7-16. *The SDL surface covers 3/4 of the layout height*

The Y position is changed to `height/4` in order to leave 1/4 of the layout height, at
the top, for the newly added views. This effect is shown in Figure 7-17.

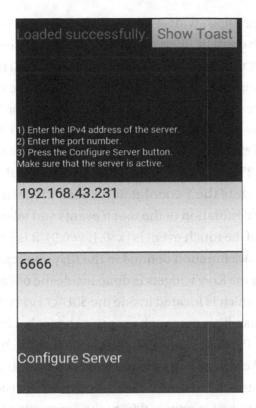

Figure 7-17. *Placing the SDL surface to the layout starting from the bottom-left corner*

Handling Touch Events of Kivy Widgets in Java

The activity layout contains the SDLSurface as a child view. This view holds all the widgets defined in the KV file of the Kivy application. In this section, we discuss how events are handled for these widgets within the Android Studio project.

In the KV file of the Python application, the four widgets are added to a BoxLayout. It has four child widgets (one Button, two TextInput, and one Label). Due to its vertical orientation, the widgets span the entire width of the BoxLayout but share the height equally. The height of the screen is divided by four and thus the height of each widget is 25% of the screen height (1/4)x1.0=0.25).

Because the (0, 0) point of the SDL Surface starts at the top-left corner, the Y coordinate of the region that the Button widget covers in the layout ranges from 0.75 to 1.0. SDL associates any touch event within that range with the button widget.

The Y range of the TextInput widget above the button goes from 0.5 to 0.75. Thus, SDL associates any touch event in this region to that widget. The same process repeats for the second TextInput widget (0.25 to 0.5) and the Label (0.0 to 0.25).

After changing the height of the surface to 3/4=0.75 of the screen height, the Y range of the widgets will change. For example, the Label widget Y coordinate will start at 0.0 but ends at 1/4 of the new surface height, which is 0.75, which is 0.25x0.75=0.1875. The Y coordinate of the first TextInput will start at 0.1875 and end at 0.375. Thus, the screen Y range from 0.1875 to 0.25 is now associated with the TextInput widget.

By calculating the range of the Y coordinate for each widget in the SDL surface, SDL can easily determine the destination of the touch events and take the proper action. For example, if the position of the touch event is (x=0.1, y=0.9), it is a touch on the Button widget and thus the callback function defined in the Kivy application will be called. Touch event handling for the Kivy widgets is done inside the onTouch() callback method of the SDLSurface class, which is located inside the SDLActivity Java file. The part of this method responsible for handling the touch events is shown in Listing 7-35.

This method accepts the target view and the event. The touch event might have multiple actions to be taken including pressing, releasing, motion, and more. For that reason, a switch block filters the action to make the proper decision.

First, the action of the event is returned from the event using the getActionMasked() method in the action integer variable. For example, when the touch is pressed, the returned integer is 0. When it is released, the integer is 1.

The numbers associated with the actions are defined as static variables in the MotionEvent class header. For example, the ACTION_DOWN variable holds the value 0 and the ACTION_UP variable has the value 1.

Listing 7-35. Handling the Touch Events of Kivy Widgets Inside Android Studio

```java
public boolean onTouch(View v, MotionEvent event) {
    ...
        int action = event.getActionMasked();
        switch(action) {
            case MotionEvent.ACTION_MOVE:
                for (i = 0; i < pointerCount; i++) {
                    pointerFingerId = event.getPointerId(i);
                    x = event.getX(i) / mWidth;
                    y = event.getY(i) / mHeight;
```

```
                            p = event.getPressure(i);
                            SDLActivity.onNativeTouch(touchDevId, pointerFingerId,
                            action, x, y, p);
                    }
                    break;

            case MotionEvent.ACTION_UP:
            case MotionEvent.ACTION_DOWN:
                    i = 0;
            case MotionEvent.ACTION_POINTER_UP:
            case MotionEvent.ACTION_POINTER_DOWN:
                    if (i == -1) {
                            i = event.getActionIndex();
                    }

                    pointerFingerId = event.getPointerId(i);
                    x = event.getX(i) / mWidth;
                    y = event.getY(i) / mHeight;
                    p = event.getPressure(i);
                    SDLActivity.onNativeTouch(touchDevId, pointerFingerId,
                    action, x, y, p);
                    break;

            case MotionEvent.ACTION_CANCEL:
                    for (i = 0; i < pointerCount; i++) {
                            pointerFingerId = event.getPointerId(i);
                            x = event.getX(i) / mWidth;
                            y = event.getY(i) / mHeight;
                            p = event.getPressure(i);
                            SDLActivity.onNativeTouch(touchDevId, pointerFingerId,
                            MotionEvent.ACTION_UP, x, y, p);
                    }
                    break;

            default:
                    break;
        ...
}
```

Inside an action such as `ACTION_POINTER_DOWN`, the x and y coordinates of the touch event are returned using the `getX()` and `getY()` methods. In order to know its relative position to the SDL surface size, they are divided by the width and height of the surface stored in the `mWidth` and `mHeight` variables. Remember that these variables are set equal to the width and height of the surface inside the `surfaceChanged()` callback method. Finally, the `onNativeTouch()` native function is called for handling the action according to the instructions written in Python. This method accepts the touch coordinates to decide which widget receives the event.

For the actions `ACTION_UP`, `ACTION_DOWN`, `ACTION_POINTER_UP`, and `ACTION_POINTER_DOWN`, the same code is executed. This is why there is no `break` statement for the cases associated with the actions `ACTION_UP`, `ACTION_DOWN`, and `ACTION_POINTER_UP`. If the action number is 1, for example, the `switch` block enters the case associated with `ACTION_UP`. Because this case does not have a `break` statement, the next case is executed, which is `ACTION_DOWN`. Also, this case does not have a `break` statement and thus the next case enters.

Similarly, the next case associated with `ACTION_POINTER_UP` does not have a `break` statement and thus the case associated with `ACTION_POINTER_DOWN` enters. Because this case includes a `break` statement at its end, its code is executed and the switch breaks.

Having reached this point, we have discussed the most important parts of the Java project created using Buildozer from the Kivy project. This helps us to understand how things work and modify the project to add Java components that work with the project.

Remember that one target of working on the Kivy project in Android Studio is that the complexity of implementing some actions within Python increases even with Pyjnius. In the next section, a Kivy button touch event is handled in Java.

Handling Kivy Button Click in Android Studio

Python-for-Android supports a list of libraries that can be packaged within an Android APK file. The list can be found at `https://github.com/kivy/python-for-android/tree/master/pythonforandroid/recipes`. Examples of these libraries are Plyer, requests, OpenCV, NumPy, and Pyjnius. Some of these libraries are well supported and work fine, but others do not. Generally, building an APK from a Kivy project that uses the image processing libraries such as NumPy and OpenCV is a bit challenging and expected to fail. Some issues in GitHub are not yet fixed, although the Kivy developers are working to solve them.

We can benefit from editing the Kivy project within Android Studio by implementing the components that are not easily implemented in Python.

In this section, a Kivy application will be created in which the user selects an image file. There is `Button` widget that's unhandled within Python. Inside Android Studio, the touch event of that button is handled within the `onTouch()` callback method discussed previously. When the button is touched, the selected image is processed using the OpenCV library. We will start by discussing the Kivy application and then downloading and importing OpenCV in order to handle the button touch event.

Kivy Application

The KV file of the application is shown in Listing 7-36. There is a new widget named `FileChooserIconView`, which displays the filesystem of the device. The `path` property specifies the path in which the files and folders are displayed. It is set to the current directory. The `dirselect` determines whether the user can select directories or not. It is set to False to prevent the user from selecting directories. The `multiselect` property is set to `False` to avoid selecting multiple files. We just need to select a single image file. When the user makes a selection, the `on_selection` event is fired. It is handled within the Python file using a callback function named `load_file()`.

After selecting the image file, the image is displayed on the `Image` widget by setting its source property to the selected file directory. A `Label` widget displays informational messages such as indicating whether the user selected an image or not.

The `Button` widget is not handled within Python. It will be handled within the Android Studio project.

Listing 7-36. KV File for a Kivy Application with a Button to be Handled Within Android Studio

```
BoxLayout:
    orientation: "vertical"
    Image:
        id: img
    FileChooserIconView:
        id: fileChooser
        path: "."
        dirselect: False
        on_selection: app.load_file()
```

```
Label:
    font_size: 20
    text_size: self.width, None
    id: selectedFile
    text: "Please choose an image (png/jpg)."
Button:
    font_size: 20
    text: "Process image in Java"
```

Figure 7-18 shows the result after running the application. Note that there is no image displayed on the Image widget and thus its color is just a white region.

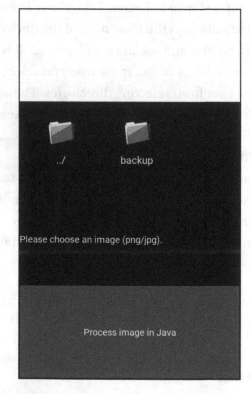

Figure 7-18. *Window of the application created in Listing 7-36*

The Python file is shown in Listing 7-37. The application class is named AndroidStudioApp and thus the KV filename must be androidstudio.kv in order to implicitly link it to the application. The class implements the load_file() callback function. This function is called even when a directory is selected.

Listing 7-37. Python File of the Application in which an Image File Is Loaded to be Processed in Android Studio Using OpenCV

```python
import kivy.app
import shutil
import os

class AndroidStudioApp(kivy.app.App):

    def load_file(self):
        fileChooser = self.root.ids['fileChooser']
        selectedFile = self.root.ids['selectedFile']
        img = self.root.ids['img']

        file_dir = fileChooser.selection[0]
        file_path, file_ext = os.path.splitext(file_dir)
        file_name = file_path.split('/')[-1]
        print(file_name, file_ext)

        if file_ext in ['.png', '.jpg']:
            img.source = file_dir
            try:
                shutil.copy(file_dir, '.')
                selectedFile.text = "**Copied**\nSelected File Directory : " + file_dir
                os.rename(".join([file_name, file_ext]), ".join(["processImageJavaOpenCV", file_ext]))
            except:
                selectedFile.text = "**Not Copied - File Already Exists**\nSelected File Directory : " + file_dir
                print("File already exists")
                os.rename(".join([file_name, file_ext]), ".join(["processImageJavaOpenCV", file_ext]))
```

```
        else:
            selectedFile.text = "The file you chose is not an image.
            Please choose an image (png/jpg)."

app = AndroidStudioApp()
app.run()
```

The function starts by accessing the widgets defined in the KV file. The file chooser is returned to the `fileChooser` variable. Even when disabling multi-selection, the selected file directory is returned into a list that can be accessed using the `selection` property. Index 0 returns the directory of the selected file as a string into the `file_dir` variable.

Because there is no guarantee that the selected file is an image, we need to verify that it is an image by checking its extension. Using the `splitext()` function in the `os.path` module, the file directory is separated into an extension and a path. The extension is returned to the `file_ext` variable and the path is returned to the `file_path` variable. The file path can be as follows:

```
NewApp/kivy/img
```

The `file_path` is also split using the `split()` function in order to return the filename to the `file_name` variable. The separator is `/`. The result is a list as shown here. The filename is the last element that can be indexed using `-1`. The filename is used later to rename the selected file.

```
['NewApp', 'kivy', 'img']
```

The selected file extension is compared to a list with the target extensions, which are `.png` and `.jpg`. If the list is equal to either, we are sure that the selected file is an image. Thus, the source property of the `Image` widget is set to the image directory.

If the extension is not `.png` or `.jpg`, a message is printed on the `Label` widget to inform the user of the incorrect selection.

The image is copied from its original path to the current directory using the `copy()` function within the `shutil` library. This copy throws an exception if the file already exists in the target directory. This is why it is written inside the `try` block. Inside the `except` block, the `Label` widget indicates that the image exists in the current directory, as shown in Figure 7-19.

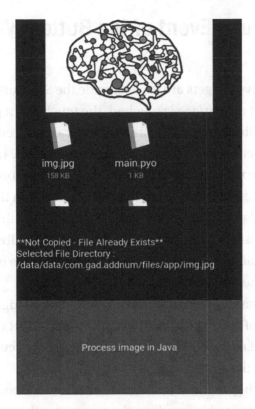

Figure 7-19. *Selecting an image file that exists in the current directory*

Inside the Java file, we need a way to indicate the name of the image file. We can standardize the process by setting the name of the image to be processed inside the Java file to a fixed name. The selected name is `processImageJavaOpenCV`. Using the `os.rename()` function, the old and new filenames are specified. Using the `join()` function, the filename and its extension are concatenated. This is done within the `try` and `except` blocks.

After making sure the Python application works well, we need to build the Android project using Buildozer. The produced project will be opened in Android Studio, as is done in the previous project. Remember to set the compile SDK and build tool versions to whatever your PC requires.

Detecting the Touch Event of the Button Widget Inside Java

Remember that the all Kivy widgets are drawn inside the SDL surface defined within the SDLActivity class. When the surface is touched, the touch event position is returned within the onTouch() callback method. This position is compared against the regions of each Kivy widget. The event is associated with the widget where the touch occurred in its region. In this section, we will detect the region of the Kivy button.

Similar to the previous example within the previous section titled "Handling Touch Events of Kivy Widgets in Java," the widgets are added within the KV file inside a BoxLayout with vertical orientation. Thus, the screen height is divided equally across all child widgets. Note that the screen and the BoxLayout can be used interchangeably because the layout fills the screen.

Because the screen has four widgets, the height for each widget is 25% of the screen height. Thus, the values of the Y coordinate of the Button widget relative to the screen height start at Y=0.75 and end at Y=1.0. Any touch event that occurs in this region of the screen is associated with the Button widget.

The modified onTouch() method that handles the button touch event is shown in Listing 7-38. If we are interested in handling the event after the touch is released, the code will be written inside the case associated with ACTION_UP.

Using an if statement, the Y coordinate of the touch position is compared against the range of Y values of the button. If the touch falls within this range, the required operations for handling the touch event are added within the if block. Currently, there is just a toast message shown. Later, after linking OpenCV with the project, some operations will be applied over the selected image.

Because there is no action associated with the widgets inside the Kivy project, we do not need to call the onNativeTouch(). It is enough to return the X and Y coordinates of the touch event to determine the target widget.

There is a break statement at the end of the if block to avoid executing more cases. If the touch is outside the range, the normal procedure followed previously will be applied by executing the code within the case associated with ACTION_POINTER_DOWN.

Listing 7-38. Handling the Touch Event of the Kivy Button Widget Within the onTouch() Method of the Android Studio Project

```
public boolean onTouch(View v, MotionEvent event) {
    ...
        switch(action) {
            case MotionEvent.ACTION_MOVE:
    ...

            case MotionEvent.ACTION_UP:
                if (i == -1) {
                    i = event.getActionIndex();
                }
                x = event.getX(i) / mWidth;
                y = event.getY(i) / mHeight;
                if (y >= 0.75){
                    Toast.makeText(this.getContext(), "Button Clicked",
                    Toast.LENGTH_LONG).show();
                break;
                }
            case MotionEvent.ACTION_DOWN:
    ...

            case MotionEvent.ACTION_POINTER_UP:
            case MotionEvent.ACTION_POINTER_DOWN:
    ...

            case MotionEvent.ACTION_CANCEL:
    ...
}
```

After building and running the Android project with the modified onTouch() method and then pressing the button, the result is shown in Figure 7-20. The next step is to link the project with OpenCV in order to process the image.

Figure 7-20. *Handling the touch event of the Kivy Button widget within Android Studio*

Importing OpenCV in Android Studio

OpenCV supports a release that can work in Android devices. The latest release for Android at this time is 3.4.4. It can be downloaded at `https://sourceforge.net/projects/opencvlibrary/files/3.4.4/opencv-3.4.4-android-sdk.zip`.

OpenCV will be imported in Android Studio as a module in order to access it within the Java code. In the File menu of Android Studio, go to New and then choose Import Module. This opens a window asking for the module source directory. You can either copy and paste the directory or navigate within your file system to locate it. If the compressed file is extracted in a folder named `opencv-3.4.4-android-sdk`, the directory to be copied and pasted is `\opencv-3.4.4-android-sdk\OpenCV-android-sdk\sdk\java`. When you enter the directory, the module name will be automatically detected as `openCVLibrary344`.

There might be an error after importing the module due to the inconsistency in the versions of the SDK and the build tools. There is a `build.gradle` file within the imported module. You can change the compile, minimum, and target SDK version to match the ones you have on your machine. This is in addition to the build tools version. This is similar to what we did previously within the build.gradle file of the project itself.

The next step is to add the imported library as a dependency in the Android project. Choose the File menu and select the Project Structure menu item. This opens a window in which the project name exists under the Modules section, according to Figure 7-21. The OpenCV library is also listed.

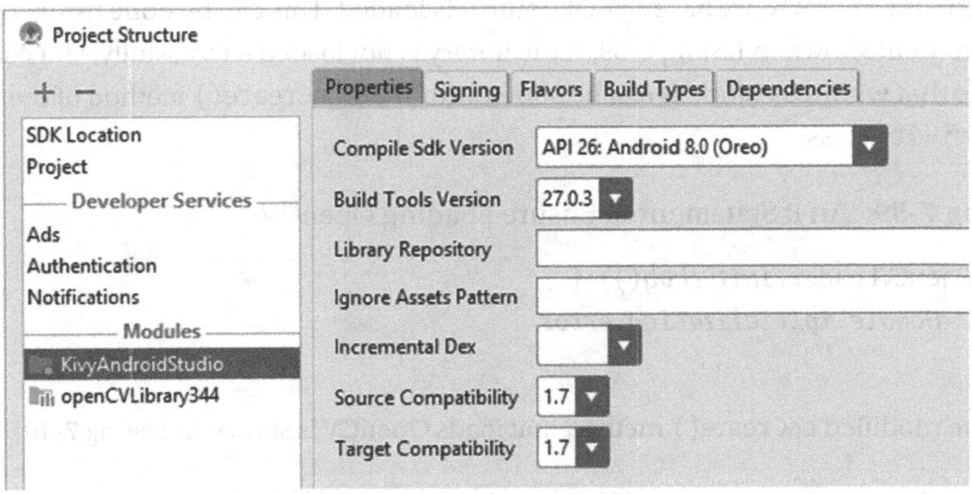

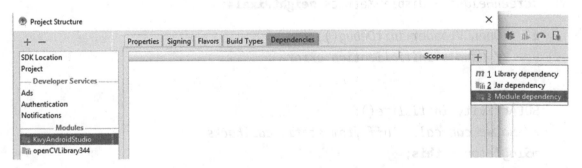

Figure 7-21. *Structure of the Android Studio project*

Go to the Dependencies tab and then click the Add button. Then choose Module Dependency, according to Figure 7-22. This opens another window in which OpenCV is listed. Select it and then click OK to add it as a dependency. Click OK again to close the Project Structure window.

Figure 7-22. *Adding OpenCV as a dependency library in the Android Studio project*

The final step is to copy a folder named libs in this path under OpenCV \OpenCV\ opencv-3.4.4-android-sdk\OpenCV-android-sdk\sdk\native. After changing the folder name to jniLibs, paste it into the \KivyAndroidStudio\src\main directory under the Android Studio project. After this step, OpenCV is ready to use.

Applying Canny Filter Over the Selected Image

Before using OpenCV, we have to make sure it is loaded. This can be done using the if statement shown in Listing 7-39. If the library is not loaded successfully, you can handle that within its block. It can be added within the onCreate() method of the SDLActivity class.

Listing 7-39. An if Statement to Ensure Loading OpenCV

```
if (!OpenCVLoader.initDebug()) {
    // Handle initialization error
}
```

The modified onCreate() method that loads OpenCV is shown in Listing 7-40.

Listing 7-40. Loading OpenCV Within the onCreate() Method

```
protected void onCreate(Bundle savedInstanceState) {
    super.onCreate(savedInstanceState);
    DisplayMetrics displayMetrics = new DisplayMetrics();
    getWindowManager().getDefaultDisplay().getMetrics(displayMetrics);
    screenWidth = displayMetrics.widthPixels;
    screenHeight = displayMetrics.heightPixels;

    if (!OpenCVLoader.initDebug()) {
        // Handle initialization error
    }

    SDLActivity.initialize();
    // So we can call stuff from static callbacks
    mSingleton = this;
}
```

After OpenCV is loaded, we can use it. The code in Listing 7-41 reads the selected image into a variable named inputImage. It assumes that the image extension is .jpg. You can add a little modification to support.png and .jpg.

The Canny edge detector is applied over that image. The resultant image of Canny is saved into another variable named outputImage. That image is saved as a .jpg file.

An ImageView is added to the SDL activity layout, which displays the saved image. The ImageView is added to the region associated with the file chooser widget that's defined in the KV file.

In order to position ImageView correctly on the screen, the screen width and height are required. There are two static integer variables named screenWidth and screenHeight defined in the SDLActivity class header. Inside the onCreate() method, the screen width and height are calculated and assigned to them.

Listing 7-41. Using OpenCV to Apply the Canny Edge Detector to the Loaded Image

```
Toast.makeText(this.getContext(), "Canny is being processed...",
Toast.LENGTH_LONG).show();
Mat inputImage = Imgcodecs.imread("processImageJavaOpenCV.jpg");
Mat outputImage = inputImage.clone();
Imgproc.Canny(inputImage, outputImage, 100, 300, 3, true);
Imgcodecs.imwrite("RESULTImageJavaOpenCV.jpg", outputImage);
File cannyImage = new File("RESULTImageJavaOpenCV.jpg");
ImageView imageView = new ImageView(this.getContext());

if(cannyImage.exists()){
    Bitmap myBitmap = BitmapFactory.decodeFile(cannyImage.
    getAbsolutePath());
    imageView.setImageBitmap(myBitmap);
    SDLActivity.mLayout.addView(imageView, new AbsoluteLayout.LayoutParams(
    SDLActivity.screenWidth, SDLActivity.screenHeight/4, 0, SDLActivity.
    screenHeight/4));
}
```

The code in Listing 7-41 can be added to the onTouch() method inside the case associated with ACTION_UP. This executes the code after pressing the Button widget defined in Kivy. Figure 7-23 shows the result after we press the button.

Figure 7-23. *OpenCV is used to apply the Canny edge detector to the loaded image*

Summary

Python alone is not able to build rich Android apps and this is why this chapter discussed enriching the Android apps created in Python using Kivy in different ways. The first library we discussed was Plyer, which allows us to access Android features using native Python code. Because many features are not implemented yet in Plyer, another library called Pyjnius was discussed, which reflects the Java classes within Python to access Java features. Unfortunately, Pyjnius adds complexity to building Java features and it is not the recommended way to enrich the Android apps. The most desirable way is to edit the Android Studio project exported using Buildozer. This way, there is no limit in what can be done in the Android apps. Anything not available in Python could be added within Java after importing the exported Android Studio project.

This chapter started discussing the project by highlighting the important files within it before making any edits. These files include the main Java activity, the manifest file, the strings resource file, and more. After that, these files were edited to add more functionality to the Android app. This included editing the loading screen layout, adding Android views to the main activity, handling click actions to both widgets, displaying toast messages, and more. This chapter also imported OpenCV within the Android Studio project in order to apply a simple filter over an image.

At the end, this chapter proved that it is possible to create rich Android apps in Python even if Python is limited in building Android apps. This is accomplished by editing the exported Android Studio project and making the necessary changes to add more functionality.

APPENDIX

Source Code

This appendix references the initial release source code for this book. For updates that will happen occasionally to the code base as necessary, please check the GitHub by going to https://www.apress.com/9781484250303.

Chapter 6: CoinTex Complete Game

KV File

```
ScreenManager:
    MainScreen:
    Level1:
    Level2:
    Level3:
    Level4:
    Level5:
    Level6:
    Level7:
    Level8:
    Level9:
    Level10:
    Level11:
    Level12:
    Level13:
    Level14:
    Level15:
    Level16:
    Level17:
```

© Ahmed Fawzy Mohamed Gad 2019

A. F. M. Gad, *Building Android Apps in Python Using Kivy with Android Studio*,
https://doi.org/10.1007/978-1-4842-5031-0

```
    Level18:
    Level19:
    Level20:
    Level21:
    Level22:
    Level23:
    Level24:
    AboutUs:
    AllLevelsCompleted:

<AboutUs>:
    name: "aboutus"
    GridLayout:
        rows: 2
        canvas.before:
            Color:
                rgba: (0.6, 0.6, 0.6, 0.8)
            Rectangle:
                pos: self.pos
                size: self.size
                source: "bg_cointex.png"
        Label:
            size_hint_y: 0.9
            text_size: self.size
            font_size: 20
            halign: 'left'
            valign: 'top'
            text: "CoinTex version 1.2 is developed by Ahmed Fawzy Gad who
            is interested in machine learning, deep learning, computer
            vision, and Python. The game is completely developed in
            Python using the Kivy cross-platform framework.\n\nFeel free
            to contact me for any questions or recommendations:\nE-mail:
            ahmed.f.gad@gmail.com\nFacebook: https://www.facebook.com/
            ahmed.f.gadd\nLinkedIn: https://www.linkedin.com/in/ahmedfgad\
            nTowardsDataScience: https://towardsdatascience.com/@ahmedfgad \
            nResearchGate: https://www.researchgate.net/profile/Ahmed_Gad13"
```

```
    ImageButton:
        size_hint_y: 0.1
        source: "other-images/Main-Screen.png"
        on_release: app.root.current="main"

<AllLevelsCompleted>:
    name: "alllevelscompleted"
    GridLayout:
        rows: 2
        canvas.before:
            Color:
                rgba: (0.6, 0.6, 0.6, 0.8)
            Rectangle:
                pos: self.pos
                size: self.size
                source: "bg_cointex.png"
        Image:
            size_hint_y: 0.9
            source: "other-images/Congratulations-Passed.png"
        ImageButton:
            size_hint_y: 0.1
            source: "other-images/Main-Screen.png"
            on_release: app.root.current="main"

<MainScreen>:
    name: "main"
    on_enter: app.main_screen_on_enter()
    on_leave: app.main_screen_on_leave()
    BoxLayout:
        orientation: "vertical"
        BoxLayout:
            size_hint_y: 0.08
            ImageButton:
                source: "./other-images/About-Us.png"
                on_release: app.root.current="aboutus"
```

```
GridLayout:
    id: lvls_imagebuttons
    cols: 4
    spacing: (5,5)
    size_hint_y: 0.92
    canvas.before:
        Rectangle:
            pos: self.pos
            size: self.size
            source: "bg_cointex.png"
    ImageButton:
        source: "levels-images/1.png"
        on_release: app.root.current="level1"
    ImageButton:
        source: "levels-images/2.png"
        on_release: app.root.current="level2"
    ImageButton:
        source: "levels-images/3.png"
        on_release: app.root.current="level3"
    ImageButton:
        source: "levels-images/4.png"
        on_release: app.root.current="level4"
    ImageButton:
        source: "levels-images/5.png"
        on_release: app.root.current="level5"
    ImageButton:
        source: "levels-images/6.png"
        on_release: app.root.current="level6"
    ImageButton:
        source: "levels-images/7.png"
        on_release: app.root.current="level7"
    ImageButton:
        source: "levels-images/8.png"
        on_release: app.root.current="level8"
```

```
ImageButton:
    source: "levels-images/9.png"
    on_release: app.root.current="level9"
ImageButton:
    source: "levels-images/10.png"
    on_release: app.root.current="level10"
ImageButton:
    source: "levels-images/11.png"
    on_release: app.root.current="level11"
ImageButton:
    source: "levels-images/12.png"
    on_release: app.root.current="level12"
ImageButton:
    source: "levels-images/13.png"
    on_release: app.root.current="level13"
ImageButton:
    source: "levels-images/14.png"
    on_release: app.root.current="level14"
ImageButton:
    source: "levels-images/15.png"
    on_release: app.root.current="level15"
ImageButton:
    source: "levels-images/16.png"
    on_release: app.root.current="level16"
ImageButton:
    source: "levels-images/17.png"
    on_release: app.root.current="level17"
ImageButton:
    source: "levels-images/18.png"
    on_release: app.root.current="level18"
ImageButton:
    source: "levels-images/19.png"
    on_release: app.root.current="level19"
ImageButton:
    source: "levels-images/20.png"
    on_release: app.root.current="level20"
```

```
            ImageButton:
                source: "levels-images/21.png"
                on_release: app.root.current="level21"
            ImageButton:
                source: "levels-images/22.png"
                on_release: app.root.current="level22"
            ImageButton:
                source: "levels-images/23.png"
                on_release: app.root.current="level23"
            ImageButton:
                source: "levels-images/24.png"
                on_release: app.root.current="level24"

<Level1>:
    name: "level1"
    on_pre_enter: app.screen_on_pre_enter(1)
    on_pre_leave: app.screen_on_pre_leave(1)
    on_enter: app.screen_on_enter(1)
    FloatLayout:
        id: layout_lvl1
        on_touch_down: app.touch_down_handler(1, args)
        canvas.before:
            Rectangle:
                pos: self.pos
                size: self.size
                source: "levels-bg/bg_lvl1.jpg"
        NumCollectedCoins:
            id: num_coins_collected_lvl1
        LevelNumber
            id: level_number_lvl1
        RemainingLifePercent
            id: remaining_life_percent_lvl1
        Monster:
            id: monster1_image_lvl1
            monst_anim_duration_low: 3.5
            monst_anim_duration_high: 4.5
```

```
        Character:
            id: character_image_lvl1
<Level2>:
    name: "level2"
    on_pre_enter: app.screen_on_pre_enter(2)
    on_pre_leave: app.screen_on_pre_leave(2)
    on_enter: app.screen_on_enter(2)
    FloatLayout:
        id: layout_lvl2
        on_touch_down: app.touch_down_handler(2, args)
        canvas.before:
            Rectangle:
                pos: self.pos
                size: self.size
                source: "levels-bg/bg_lvl2.jpg"
        NumCollectedCoins:
            id: num_coins_collected_lvl2
        LevelNumber
            id: level_number_lvl2
        RemainingLifePercent
            id: remaining_life_percent_lvl2
        Monster:
            id: monster1_image_lvl2
            monst_anim_duration_low: 2.5
            monst_anim_duration_high: 3.5
        Character:
            id: character_image_lvl2
<Level3>:
    name: "level3"
    on_pre_enter: app.screen_on_pre_enter(3)
    on_pre_leave: app.screen_on_pre_leave(3)
    on_enter: app.screen_on_enter(3)
    FloatLayout:
        id: layout_lvl3
        on_touch_down: app.touch_down_handler(3, args)
```

```
        canvas.before:
            Rectangle:
                pos: self.pos
                size: self.size
                source: "levels-bg/bg_lvl3.jpg"
        NumCollectedCoins:
            id: num_coins_collected_lvl3
        LevelNumber
            id: level_number_lvl3
        RemainingLifePercent
            id: remaining_life_percent_lvl3
        Monster:
            id: monster1_image_lvl3
            monst_anim_duration_low: 1.0
            monst_anim_duration_high: 1.5
        Character:
            id: character_image_lvl3

<Level4>:
    name: "level4"
    on_pre_enter: app.screen_on_pre_enter(4)
    on_pre_leave: app.screen_on_pre_leave(4)
    on_enter: app.screen_on_enter(4)
    FloatLayout:
        id: layout_lvl4
        on_touch_down: app.touch_down_handler(4, args)
        canvas.before:
            Rectangle:
                pos: self.pos
                size: self.size
                source: "levels-bg/bg_lvl4.jpg"
        NumCollectedCoins:
            id: num_coins_collected_lvl4
        LevelNumber
            id: level_number_lvl4
```

```
RemainingLifePercent
    id: remaining_life_percent_lvl4
Fire:
    id: fire1_lvl4
    fire_start_pos_hint: {'x': 1.5, 'y': 0.5}
    fire_end_pos_hint: {'x': -0.5, 'y': 0.5}
    fire_anim_duration: 2.5
Monster:
    id: monster1_image_lvl4
    monst_anim_duration_low: 2.5
    monst_anim_duration_high: 3.5
Character:
    id: character_image_lvl4

<Level5>:
    name: "level5"
    on_pre_enter: app.screen_on_pre_enter(5)
    on_pre_leave: app.screen_on_pre_leave(5)
    on_enter: app.screen_on_enter(5)
    FloatLayout:
        id: layout_lvl5
        on_touch_down: app.touch_down_handler(5, args)
        canvas.before:
            Rectangle:
                pos: self.pos
                size: self.size
                source: "levels-bg/bg_lvl5.jpg"
        NumCollectedCoins:
            id: num_coins_collected_lvl5
        LevelNumber
            id: level_number_lvl5
        RemainingLifePercent
            id: remaining_life_percent_lvl5
        Fire:
            id: fire1_lvl5
            fire_start_pos_hint: {'x': 1.5, 'y': 0.7}
```

```
            fire_end_pos_hint: {'x': -0.5, 'y': 0.7}
            fire_anim_duration: 3.0
        Fire:
            id: fire2_lvl5
            fire_start_pos_hint: {'x': 1.5, 'y': 0.3}
            fire_end_pos_hint: {'x': -0.5, 'y': 0.3}
            fire_anim_duration: 2.0
        Monster:
            id: monster1_image_lvl5
            monst_anim_duration_low: 1.0
            monst_anim_duration_high: 1.5
        Character:
            id: character_image_lvl5

<Level6>:
    name: "level6"
    on_pre_enter: app.screen_on_pre_enter(6)
    on_pre_leave: app.screen_on_pre_leave(6)
    on_enter: app.screen_on_enter(6)
    FloatLayout:
        id: layout_lvl6
        on_touch_down: app.touch_down_handler(6, args)
        canvas.before:
            Rectangle:
                pos: self.pos
                size: self.size
                source: "levels-bg/bg_lvl6.jpg"
        NumCollectedCoins:
            id: num_coins_collected_lvl6
        LevelNumber
            id: level_number_lvl6
        RemainingLifePercent
            id: remaining_life_percent_lvl6
        Fire:
            id: fire1_lvl6
            fire_start_pos_hint: {'x': 1.5, 'y': 0.8}
```

```
            fire_end_pos_hint: {'x': -0.5, 'y': 0.8}
            fire_anim_duration: 2.5
        Fire:
            id: fire2_lvl6
            fire_start_pos_hint: {'x': 1.5, 'y': 0.2}
            fire_end_pos_hint: {'x': -0.5, 'y': 0.2}
            fire_anim_duration: 2.0
        Fire:
            id: fire3_lvl6
            fire_start_pos_hint: {'x': 0.5, 'y': 1.5}
            fire_end_pos_hint: {'x': 0.5, 'y': -0.5}
            fire_anim_duration: 3.5
        Monster:
            id: monster1_image_lvl6
            monst_anim_duration_low: 2.5
            monst_anim_duration_high: 3.0
        Character:
            id: character_image_lvl6

<level7>:
    name: "level7"
    on_pre_enter: app.screen_on_pre_enter(7)
    on_pre_leave: app.screen_on_pre_leave(7)
    on_enter: app.screen_on_enter(7)
    FloatLayout:
        id: layout_lvl7
        on_touch_down: app.touch_down_handler(7, args)
        canvas.before:
            Rectangle:
                pos: self.pos
                size: self.size
                source: "levels-bg/bg_lvl7.jpg"
        NumCollectedCoins:
            id: num_coins_collected_lvl7
        LevelNumber
            id: level_number_lvl7
```

```
        RemainingLifePercent
            id: remaining_life_percent_lvl7
        Monster:
            id: monster1_image_lvl7
            monst_anim_duration_low: 1.0
            monst_anim_duration_high: 1.5
        Monster:
            id: monster2_image_lvl7
            monst_anim_duration_low: 1.5
            monst_anim_duration_high: 2.5
        Monster:
            id: monster3_image_lvl7
            monst_anim_duration_low: 1.5
            monst_anim_duration_high: 2.5
        Character:
            id: character_image_lvl7

<Level8>:
    name: "level8"
    on_pre_enter: app.screen_on_pre_enter(8)
    on_pre_leave: app.screen_on_pre_leave(8)
    on_enter: app.screen_on_enter(8)
    FloatLayout:
        id: layout_lvl8
        on_touch_down: app.touch_down_handler(8, args)
        canvas.before:
            Rectangle:
                pos: self.pos
                size: self.size
                source: "levels-bg/bg_lvl8.jpg"
        NumCollectedCoins:
            id: num_coins_collected_lvl8
        LevelNumber
            id: level_number_lvl8
        RemainingLifePercent
            id: remaining_life_percent_lvl8
```

```
    Monster:
        id: monster1_image_lvl8
        monst_anim_duration_low: 1.5
        monst_anim_duration_high: 2.5
    Monster:
        id: monster2_image_lvl8
        monst_anim_duration_low: 1.0
        monst_anim_duration_high: 1.5
    Character:
        id: character_image_lvl8

<Level9>:
    name: "level9"
    on_pre_enter: app.screen_on_pre_enter(9)
    on_pre_leave: app.screen_on_pre_leave(9)
    on_enter: app.screen_on_enter(9)
    FloatLayout:
        id: layout_lvl9
        on_touch_down: app.touch_down_handler(9, args)
        canvas.before:
            Rectangle:
                pos: self.pos
                size: self.size
                source: "levels-bg/bg_lvl9.jpg"
        NumCollectedCoins:
            id: num_coins_collected_lvl9
        LevelNumber
            id: level_number_lvl9
        RemainingLifePercent
            id: remaining_life_percent_lvl9
        Monster:
            id: monster1_image_lvl9
            monst_anim_duration_low: 2.5
            monst_anim_duration_high: 3.5
        Monster2:
            id: monster2_image_lvl9
```

```
            monst_anim_duration_low: 1.2
            monst_anim_duration_high: 1.7
        Character:
            id: character_image_lvl9

<Level10>:
    name: "level10"
    on_pre_enter: app.screen_on_pre_enter(10)
    on_pre_leave: app.screen_on_pre_leave(10)
    on_enter: app.screen_on_enter(10)
    FloatLayout:
        id: layout_lvl10
        on_touch_down: app.touch_down_handler(10, args)
        canvas.before:
            Rectangle:
                pos: self.pos
                size: self.size
                source: "levels-bg/bg_lvl10.jpg"
        NumCollectedCoins:
            id: num_coins_collected_lvl10
        LevelNumber
            id: level_number_lvl10
        RemainingLifePercent
            id: remaining_life_percent_lvl10
        Monster:
            id: monster1_image_lvl10
            monst_anim_duration_low: 1.5
            monst_anim_duration_high: 2.5
        Monster2:
            id: monster2_image_lvl10
            monst_anim_duration_low: 0.8
            monst_anim_duration_high: 1.2
        Monster2:
            id: monster3_image_lvl10
            monst_anim_duration_low: 1.0
            monst_anim_duration_high: 1.5
```

```
        Character:
            id: character_image_lvl10

<Level11>:
    name: "level11"
    on_pre_enter: app.screen_on_pre_enter(11)
    on_pre_leave: app.screen_on_pre_leave(11)
    on_enter: app.screen_on_enter(11)
    FloatLayout:
        id: layout_lvl11
        on_touch_down: app.touch_down_handler(11, args)
        canvas.before:
            Rectangle:
                pos: self.pos
                size: self.size
                source: "levels-bg/bg_lvl11.jpg"
        NumCollectedCoins:
            id: num_coins_collected_lvl11
        LevelNumber
            id: level_number_lvl11
        RemainingLifePercent
            id: remaining_life_percent_lvl11
        Fire:
            id: fire1_lvl11
            fire_start_pos_hint: {'x': 1.5, 'y': 0.5}
            fire_end_pos_hint: {'x': -0.5, 'y': 0.5}
            fire_anim_duration: 3.5
        Monster:
            id: monster1_image_lvl11
            monst_anim_duration_low: 2.5
            monst_anim_duration_high: 3.5
        Monster2:
            id: monster2_image_lvl11
            monst_anim_duration_low: 2.0
            monst_anim_duration_high: 2.5
```

```
        Character:
            id: character_image_lvl11

<Level12>:
    name: "level12"
    on_pre_enter: app.screen_on_pre_enter(12)
    on_pre_leave: app.screen_on_pre_leave(12)
    on_enter: app.screen_on_enter(12)
    FloatLayout:
        id: layout_lvl12
        on_touch_down: app.touch_down_handler(12, args)
        canvas.before:
            Rectangle:
                pos: self.pos
                size: self.size
                source: "levels-bg/bg_lvl12.jpg"
        NumCollectedCoins:
            id: num_coins_collected_lvl12
        LevelNumber
            id: level_number_lvl12
        RemainingLifePercent
            id: remaining_life_percent_lvl12
        Fire:
            id: fire1_lvl12
            fire_start_pos_hint: {'x': 1.5, 'y': 0.5}
            fire_end_pos_hint: {'x': -0.5, 'y': 0.5}
            fire_anim_duration: 2.5
        Monster:
            id: monster1_image_lvl12
            monst_anim_duration_low: 2.5
            monst_anim_duration_high: 3.5
        Monster2:
            id: monster2_image_lvl12
            monst_anim_duration_low: 1.5
            monst_anim_duration_high: 2.0
```

```
        Character:
            id: character_image_lvl12

<Level13>:
    name: "level13"
    on_pre_enter: app.screen_on_pre_enter(13)
    on_pre_leave: app.screen_on_pre_leave(13)
    on_enter: app.screen_on_enter(13)
    FloatLayout:
        id: layout_lvl13
        on_touch_down: app.touch_down_handler(13, args)
        canvas.before:
            Rectangle:
                pos: self.pos
                size: self.size
                source: "levels-bg/bg_lvl13.jpg"
        NumCollectedCoins:
            id: num_coins_collected_lvl13
        LevelNumber
            id: level_number_lvl13
        RemainingLifePercent
            id: remaining_life_percent_lvl13
        Fire:
            id: fire1_lvl13
            fire_start_pos_hint: {'x': 1.5, 'y': 0.5}
            fire_end_pos_hint: {'x': -0.5, 'y': 0.5}
            fire_anim_duration: 2.5
        Fire:
            id: fire2_lvl13
            fire_start_pos_hint: {'x': 0.5, 'y': 1.5}
            fire_end_pos_hint: {'x': 0.5, 'y': -0.5}
            fire_anim_duration: 3.5
        Monster:
            id: monster1_image_lvl13
            monst_anim_duration_low: 2.5
            monst_anim_duration_high: 3.5
```

```
        Monster2:
            id: monster2_image_lvl13
            monst_anim_duration_low: 2.5
            monst_anim_duration_high: 3.0
        Character:
            id: character_image_lvl13

<Level14>:
    name: "level14"
    on_pre_enter: app.screen_on_pre_enter(14)
    on_pre_leave: app.screen_on_pre_leave(14)
    on_enter: app.screen_on_enter(14)
    FloatLayout:
        id: layout_lvl14
        on_touch_down: app.touch_down_handler(14, args)
        canvas.before:
            Rectangle:
                pos: self.pos
                size: self.size
                source: "levels-bg/bg_lvl14.jpg"
        NumCollectedCoins:
            id: num_coins_collected_lvl14
        LevelNumber
            id: level_number_lvl14
        RemainingLifePercent
            id: remaining_life_percent_lvl14
        Fire:
            id: fire1_lvl14
            fire_start_pos_hint: {'x': 1.5, 'y': 0.2}
            fire_end_pos_hint: {'x': -0.5, 'y': 0.2}
            fire_anim_duration: 2.5
        Fire:
            id: fire2_lvl14
            fire_start_pos_hint: {'x': 1.5, 'y': 0.8}
            fire_end_pos_hint: {'x': -0.5, 'y': 0.8}
            fire_anim_duration: 3.5
```

```
            Fire:
                id: fire3_lvl14
                fire_start_pos_hint: {'x': -1.5, 'y': 0.5}
                fire_end_pos_hint: {'x': -0.5, 'y': 0.5}
                fire_anim_duration: 3.5
            Fire:
                id: fire4_lvl14
                fire_start_pos_hint: {'x': 0.8, 'y': 1.5}
                fire_end_pos_hint: {'x': 0.8, 'y': -0.5}
                fire_anim_duration: 2.5
            Fire:
                id: fire5_lvl14
                fire_start_pos_hint: {'x': 0.2, 'y': 1.5}
                fire_end_pos_hint: {'x': 0.2, 'y': -0.5}
                fire_anim_duration: 3.5
            Fire:
                id: fire6_lvl14
                fire_start_pos_hint: {'x': 0.5, 'y': 1.5}
                fire_end_pos_hint: {'x': 0.5, 'y': -0.5}
                fire_anim_duration: 2.5
            Character:
                id: character_image_lvl14

<Level15>:
    name: "level15"
    on_pre_enter: app.screen_on_pre_enter(15)
    on_pre_leave: app.screen_on_pre_leave(15)
    on_enter: app.screen_on_enter(15)
    FloatLayout:
        id: layout_lvl15
        on_touch_down: app.touch_down_handler(15, args)
        canvas.before:
            Rectangle:
                pos: self.pos
                size: self.size
                source: "levels-bg/bg_lvl15.jpg"
```

```
            NumCollectedCoins:
                id: num_coins_collected_lvl15
            LevelNumber
                id: level_number_lvl15
            RemainingLifePercent
                id: remaining_life_percent_lvl15
            Fire:
                id: fire1_lvl15
                fire_start_pos_hint: {'x': 1.2, 'y': 0.2}
                fire_end_pos_hint: {'x': -0.5, 'y': 0.2}
                fire_anim_duration: 2.5
            Fire:
                id: fire2_lvl15
                fire_start_pos_hint: {'x': 1.5, 'y': 0.7}
                fire_end_pos_hint: {'x': -0.5, 'y': 0.7}
                fire_anim_duration: 3.5
            Fire:
                id: fire3_lvl15
                fire_start_pos_hint: {'x': 0.5, 'y': 1.5}
                fire_end_pos_hint: {'x': 0.5, 'y': -0.5}
                fire_anim_duration: 2.5
            Monster2:
                id: monster1_image_lvl15
                monst_anim_duration_low: 3.0
                monst_anim_duration_high: 3.5
            Monster2:
                id: monster2_image_lvl15
                monst_anim_duration_low: 3.0
                monst_anim_duration_high: 3.5
            Character:
                id: character_image_lvl15

<Level16>:
    name: "level16"
    on_pre_enter: app.screen_on_pre_enter(16)
    on_pre_leave: app.screen_on_pre_leave(16)
```

```
on_enter: app.screen_on_enter(16)
FloatLayout:
    id: layout_lvl16
    on_touch_down: app.touch_down_handler(16, args)
    canvas.before:
        Rectangle:
            pos: self.pos
            size: self.size
            source: "levels-bg/bg_lvl16.jpg"
    NumCollectedCoins:
        id: num_coins_collected_lvl16
    LevelNumber
        id: level_number_lvl16
    RemainingLifePercent
        id: remaining_life_percent_lvl16
    Fire:
        id: fire1_lvl16
        fire_start_pos_hint: {'x': 1.5, 'y': 0.2}
        fire_end_pos_hint: {'x': -0.5, 'y': 0.2}
        fire_anim_duration: 2.5
    Fire:
        id: fire2_lvl16
        fire_start_pos_hint: {'x': 1.5, 'y': 0.7}
        fire_end_pos_hint: {'x': -0.5, 'y': 0.7}
        fire_anim_duration: 1.5
    Monster:
        id: monster1_image_lvl16
        monst_anim_duration_low: 2.0
        monst_anim_duration_high: 3.5
    Monster2:
        id: monster2_image_lvl16
        monst_anim_duration_low: 3.0
        monst_anim_duration_high: 3.5
    Monster2:
        id: monster3_image_lvl16
```

```
                monst_anim_duration_low: 1.0
                monst_anim_duration_high: 1.5
        Character:
            id: character_image_lvl16

<Level17>:
    name: "level17"
    on_pre_enter: app.screen_on_pre_enter(17)
    on_pre_leave: app.screen_on_pre_leave(17)
    on_enter: app.screen_on_enter(17)
    FloatLayout:
        id: layout_lvl17
        on_touch_down: app.touch_down_handler(17, args)
        canvas.before:
            Rectangle:
                pos: self.pos
                size: self.size
                source: "levels-bg/bg_lvl17.jpg"
        NumCollectedCoins:
            id: num_coins_collected_lvl17
        LevelNumber
            id: level_number_lvl17
        RemainingLifePercent
            id: remaining_life_percent_lvl17
        Fire:
            id: fire1_lvl17
            fire_start_pos_hint: {'x': 1.5, 'y': 0.3}
            fire_end_pos_hint: {'x': -1.5, 'y': 0.3}
            fire_anim_duration: 1.5
        Fire:
            id: fire2_lvl17
            fire_start_pos_hint: {'x': 1.5, 'y': 0.8}
            fire_end_pos_hint: {'x': -1.5, 'y': 0.8}
            fire_anim_duration: 1.0
```

```
    Fire:
        id: fire3_lvl17
        fire_start_pos_hint: {'x': 0.8, 'y': -1.5}
        fire_end_pos_hint: {'x': 0.8, 'y': 1.5}
        fire_anim_duration: 1.5
    Fire:
        id: fire4_lvl17
        fire_start_pos_hint: {'x': 0.3, 'y': 1.5}
        fire_end_pos_hint: {'x': 0.3, 'y': -1.5}
        fire_anim_duration: 1.0
    Character:
        id: character_image_lvl17

<Level18>:
    name: "level18"
    on_pre_enter: app.screen_on_pre_enter(18)
    on_pre_leave: app.screen_on_pre_leave(18)
    on_enter: app.screen_on_enter(18)
    FloatLayout:
        id: layout_lvl18
        on_touch_down: app.touch_down_handler(18, args)
        canvas.before:
            Rectangle:
                pos: self.pos
                size: self.size
                source: "levels-bg/bg_lvl18.jpg"
        NumCollectedCoins:
            id: num_coins_collected_lvl18
        LevelNumber
            id: level_number_lvl18
        RemainingLifePercent
            id: remaining_life_percent_lvl18
        Fire:
            id: fire1_lvl18
            fire_start_pos_hint: {'x': 0.8, 'y': 0.3}
            fire_end_pos_hint: {'x': 0.2, 'y': 0.3}
```

```
            fire_anim_duration: 1.2
        Fire:
            id: fire2_lvl18
            fire_start_pos_hint: {'x': 0.3, 'y': 0.8}
            fire_end_pos_hint: {'x': 0.7, 'y': 0.8}
            fire_anim_duration: 1.3
        Fire:
            id: fire3_lvl18
            fire_start_pos_hint: {'x': 0.8, 'y': 0.3}
            fire_end_pos_hint: {'x': 0.8, 'y': 0.7}
            fire_anim_duration: 1.2
        Fire:
            id: fire4_lvl18
            fire_start_pos_hint: {'x': 0.3, 'y': 0.8}
            fire_end_pos_hint: {'x': 0.3, 'y': 0.2}
            fire_anim_duration: 1.0
        Monster:
            id: monster1_image_lvl18
            monst_anim_duration_low: 1.0
            monst_anim_duration_high: 1.6
        Monster2:
            id: monster2_image_lvl18
            monst_anim_duration_low: 1.0
            monst_anim_duration_high: 2.0
        Monster2:
            id: monster3_image_lvl18
            monst_anim_duration_low: 1.0
            monst_anim_duration_high: 1.2
        Character:
            id: character_image_lvl18

<Level19>:
    name: "level19"
    on_pre_enter: app.screen_on_pre_enter(19)
    on_pre_leave: app.screen_on_pre_leave(19)
    on_enter: app.screen_on_enter(19)
```

```
FloatLayout:
    id: layout_lvl19
    on_touch_down: app.touch_down_handler(19, args)
    canvas.before:
        Rectangle:
            pos: self.pos
            size: self.size
            source: "levels-bg/bg_lvl19.jpg"
    NumCollectedCoins:
        id: num_coins_collected_lvl19
    LevelNumber
        id: level_number_lvl19
    RemainingLifePercent
        id: remaining_life_percent_lvl19
    Fire:
        id: fire1_lvl19
        fire_start_pos_hint: {'x': 0.8, 'y': 0.3}
        fire_end_pos_hint: {'x': 0.2, 'y': 0.3}
        fire_anim_duration: 1.2
    Fire:
        id: fire2_lvl19
        fire_start_pos_hint: {'x': 0.3, 'y': 0.8}
        fire_end_pos_hint: {'x': 0.7, 'y': 0.8}
        fire_anim_duration: 1.0
    Fire:
        id: fire3_lvl19
        fire_start_pos_hint: {'x': -1.5, 'y': 0.5}
        fire_end_pos_hint: {'x': 1.5, 'y': 0.5}
        fire_anim_duration: 1.5
    Fire:
        id: fire4_lvl19
        fire_start_pos_hint: {'x': 0.8, 'y': 0.3}
        fire_end_pos_hint: {'x': 0.8, 'y': 0.7}
        fire_anim_duration: 1.2
```

```
            Fire:
                id: fire5_lvl19
                fire_start_pos_hint: {'x': 0.3, 'y': 0.8}
                fire_end_pos_hint: {'x': 0.3, 'y': 0.2}
                fire_anim_duration: 1.3
            Fire:
                id: fire6_lvl19
                fire_start_pos_hint: {'x': 0.5, 'y': -1.5}
                fire_end_pos_hint: {'x': 0.5, 'y': 1.5}
                fire_anim_duration: 1.4
            Character:
                id: character_image_lvl19

<Level20>:
    name: "level20"
    on_pre_enter: app.screen_on_pre_enter(20)
    on_pre_leave: app.screen_on_pre_leave(20)
    on_enter: app.screen_on_enter(20)
    FloatLayout:
        id: layout_lvl20
        on_touch_down: app.touch_down_handler(20, args)
        canvas.before:
            Rectangle:
                pos: self.pos
                size: self.size
                source: "levels-bg/bg_lvl20.jpg"
        NumCollectedCoins:
            id: num_coins_collected_lvl20
        LevelNumber
            id: level_number_lvl20
        RemainingLifePercent
            id: remaining_life_percent_lvl20
        Fire:
            id: fire1_lvl20
            fire_start_pos_hint: {'x': 0.8, 'y': 0.3}
            fire_end_pos_hint: {'x': 0.2, 'y': 0.3}
```

```
    fire_anim_duration: 1.5
Fire:
    id: fire2_lvl20
    fire_start_pos_hint: {'x': 0.3, 'y': 0.8}
    fire_end_pos_hint: {'x': 0.7, 'y': 0.8}
    fire_anim_duration: 1.3
Fire:
    id: fire3_lvl20
    fire_start_pos_hint: {'x': -1.5, 'y': 0.5}
    fire_end_pos_hint: {'x': 1.5, 'y': 0.5}
    fire_anim_duration: 1.6
Fire:
    id: fire4_lvl20
    fire_start_pos_hint: {'x': 0.1, 'y': 0.3}
    fire_end_pos_hint: {'x': 0.9, 'y': 0.3}
    fire_anim_duration: 1.5
Fire:
    id: fire5_lvl20
    fire_start_pos_hint: {'x': 0.8, 'y': 0.3}
    fire_end_pos_hint: {'x': 0.8, 'y': 0.7}
    fire_anim_duration: 1.5
Fire:
    id: fire6_lvl20
    fire_start_pos_hint: {'x': 0.3, 'y': 0.8}
    fire_end_pos_hint: {'x': 0.3, 'y': 0.2}
    fire_anim_duration: 1.3
Fire:
    id: fire7_lvl20
    fire_start_pos_hint: {'x': 0.5, 'y': -1.5}
    fire_end_pos_hint: {'x': 0.5, 'y': 1.5}
    fire_anim_duration: 1.6
Fire:
    id: fire8_lvl20
    fire_start_pos_hint: {'x': 0.5, 'y': 0.1}
    fire_end_pos_hint: {'x': 0.5, 'y': 0.9}
    fire_anim_duration: 1.3
```

```
        Character:
            id: character_image_lvl20

<Level21>:
    name: "level21"
    on_pre_enter: app.screen_on_pre_enter(21)
    on_pre_leave: app.screen_on_pre_leave(21)
    on_enter: app.screen_on_enter(21)
    FloatLayout:
        id: layout_lvl21
        on_touch_down: app.touch_down_handler(21, args)
        canvas.before:
            Rectangle:
                pos: self.pos
                size: self.size
                source: "levels-bg/bg_lvl1.jpg"
        NumCollectedCoins:
            id: num_coins_collected_lvl21
        LevelNumber
            id: level_number_lvl21
        RemainingLifePercent
            id: remaining_life_percent_lvl21
        Fire:
            id: fire1_lvl21
            fire_start_pos_hint: {'x': 0.8, 'y': 0.3}
            fire_end_pos_hint: {'x': 0.2, 'y': 0.3}
            fire_anim_duration: 1.8
        Fire:
            id: fire2_lvl21
            fire_start_pos_hint: {'x': 0.3, 'y': 0.8}
            fire_end_pos_hint: {'x': 0.7, 'y': 0.8}
            fire_anim_duration: 1.5
        Fire:
            id: fire3_lvl21
            fire_start_pos_hint: {'x': -1.5, 'y': 0.5}
            fire_end_pos_hint: {'x': 1.5, 'y': 0.5}
```

```
            fire_anim_duration: 2.0
        Fire:
            id: fire4_lvl21
            fire_start_pos_hint: {'x': 0.1, 'y': 0.3}
            fire_end_pos_hint: {'x': 0.9, 'y': 0.3}
            fire_anim_duration: 2.5
        Monster:
            id: monster1_image_lvl21
            monst_anim_duration_low: 1.0
            monst_anim_duration_high: 1.3
        Monster2:
            id: monster2_image_lvl21
            monst_anim_duration_low: 1.0
            monst_anim_duration_high: 1.5
        Character:
            id: character_image_lvl21

<Level22>:
    name: "level22"
    on_pre_enter: app.screen_on_pre_enter(22)
    on_pre_leave: app.screen_on_pre_leave(22)
    on_enter: app.screen_on_enter(22)
    FloatLayout:
        id: layout_lvl22
        on_touch_down: app.touch_down_handler(22, args)
        canvas.before:
            Rectangle:
                pos: self.pos
                size: self.size
                source: "levels-bg/bg_lvl2.jpg"
        NumCollectedCoins:
            id: num_coins_collected_lvl22
        LevelNumber
            id: level_number_lvl22
        RemainingLifePercent
            id: remaining_life_percent_lvl22
```

```
        Fire:
            id: fire1_lvl22
            fire_start_pos_hint: {'x': 0.8, 'y': 0.3}
            fire_end_pos_hint: {'x': 0.2, 'y': 0.3}
            fire_anim_duration: 1.9
        Fire:
            id: fire2_lvl22
            fire_start_pos_hint: {'x': 0.3, 'y': 0.8}
            fire_end_pos_hint: {'x': 0.7, 'y': 0.8}
            fire_anim_duration: 2.0
        Fire:
            id: fire3_lvl22
            fire_start_pos_hint: {'x': -1.5, 'y': 0.5}
            fire_end_pos_hint: {'x': 1.5, 'y': 0.5}
            fire_anim_duration: 2.0
        Fire:
            id: fire4_lvl22
            fire_start_pos_hint: {'x': 0.1, 'y': 0.3}
            fire_end_pos_hint: {'x': 0.9, 'y': 0.3}
            fire_anim_duration: 2.5
        Monster:
            id: monster1_image_lvl22
            monst_anim_duration_low: 1.0
            monst_anim_duration_high: 1.8
        Monster2:
            id: monster2_image_lvl22
            monst_anim_duration_low: 1.0
            monst_anim_duration_high: 2.0
        Character:
            id: character_image_lvl22

<Level23>:
    name: "level23"
    on_pre_enter: app.screen_on_pre_enter(23)
    on_pre_leave: app.screen_on_pre_leave(23)
    on_enter: app.screen_on_enter(23)
```

```
FloatLayout:
    id: layout_lvl23
    on_touch_down: app.touch_down_handler(23, args)
    canvas.before:
        Rectangle:
            pos: self.pos
            size: self.size
            source: "levels-bg/bg_lvl3.jpg"
    NumCollectedCoins:
        id: num_coins_collected_lvl23
    LevelNumber
        id: level_number_lvl23
    RemainingLifePercent
        id: remaining_life_percent_lvl23
    Fire:
        id: fire1_lvl23
        fire_start_pos_hint: {'x': -1.5, 'y': 0.3}
        fire_end_pos_hint: {'x': -0.5, 'y': 0.3}
        fire_anim_duration: 1.8
    Fire:
        id: fire2_lvl23
        fire_start_pos_hint: {'x': -1.5, 'y': 0.8}
        fire_end_pos_hint: {'x': -0.5, 'y': 0.8}
        fire_anim_duration: 2.0
    Monster:
        id: monster1_image_lvl23
        monst_anim_duration_low: 1.5
        monst_anim_duration_high: 2.0
    Monster2:
        id: monster2_image_lvl23
        monst_anim_duration_low: 1.5
        monst_anim_duration_high: 2.5
    Character:
        id: character_image_lvl23
```

```
<Level24>:
    name: "level24"
    on_pre_enter: app.screen_on_pre_enter(24)
    on_pre_leave: app.screen_on_pre_leave(24)
    on_enter: app.screen_on_enter(24)
    FloatLayout:
        id: layout_lvl24
        on_touch_down: app.touch_down_handler(24, args)
        canvas.before:
            Rectangle:
                pos: self.pos
                size: self.size
                source: "levels-bg/bg_lvl4.jpg"
        NumCollectedCoins:
            id: num_coins_collected_lvl24
        LevelNumber
            id: level_number_lvl24
        RemainingLifePercent
            id: remaining_life_percent_lvl24
        Fire:
            id: fire1_lvl24
            fire_start_pos_hint: {'x': 0.8, 'y': 0.3}
            fire_end_pos_hint: {'x': 0.2, 'y': 0.3}
            fire_anim_duration: 1.8
        Fire:
            id: fire2_lvl24
            fire_start_pos_hint: {'x': 0.3, 'y': 0.8}
            fire_end_pos_hint: {'x': 0.7, 'y': 0.8}
            fire_anim_duration: 1.5
        Monster:
            id: monster1_image_lvl24
            monst_anim_duration_low: 1.5
            monst_anim_duration_high: 1.7
        Monster2:
            id: monster2_image_lvl24
```

```
            monst_anim_duration_low: 1.8
            monst_anim_duration_high: 2.5
        Monster:
            id: monster3_image_lvl24
            monst_anim_duration_low: 1.5
            monst_anim_duration_high: 2.0
        Character:
            id: character_image_lvl24

<NumCollectedCoins@Label>:
    size_hint: (0.1, 0.02)
    pos_hint: {'x': 0.026, 'y': 0.97}
    text: "Coins 0"
    font_size: 20

<LevelNumber@Label>:
    size_hint: (0.1, 0.02)
    pos_hint: {'x': 0.15, 'y': 0.97}
    text: "Level"
    font_size: 20

<RemainingLifePercent@Label>:
    remaining_life_size_hint_x: 0.2
    size_hint: (0.2, 0.022)
    pos_hint: {'x': 0.3, 'y': 0.97}
    font_size: 20
    canvas:
        Color:
            rgb: (1, 0, 0)
        Rectangle:
            pos: self.pos
            size: self.size

<Monster@Image>:
    size_hint: (0.15, 0.15)
    pos_hint: {'x': 0.8, 'y': 0.8}
    source: "10.png"
```

```
    im_num: 10
    start_im_num: 10
    end_im_num: 17
    allow_stretch: True
    on_im_num: app.change_monst_im(self)
    on_pos_hint: app.monst_pos_hint(self)

<Monster2@Image>:
    size_hint: (0.15, 0.15)
    pos_hint: {'x': 0.8, 'y': 0.8}
    source: "21.png"
    im_num: 21
    start_im_num: 21
    end_im_num: 29
    allow_stretch: True
    on_im_num: app.change_monst_im(self)
    on_pos_hint: app.monst_pos_hint(self)

<Character@Image>:
    size_hint: (0.15, 0.15)
    pos_hint: {'x': 0.0, 'y': 0.0}
    source: "0.png"
    im_num: 0
    start_im_num: 0
    end_im_num: 7
    dead_start_im_num: 91
    dead_end_im_num: 95
    allow_stretch: True
    on_im_num: app.change_char_im(self)
    on_pos_hint: app.char_pos_hint(self)

<Fire@Label>:
    pos_hint: {'x': 1.1, 'y': 1.1}
    on_pos_hint: app.fire_pos_hint(self)
    size_hint: (0.03, 0.03)
```

```
    canvas:
        Rectangle:
            pos: self.pos
            size: self.size
            source: "fire.png"

<ImageButton>:
    canvas.before:
        Color:
            rgba: (0.6, 0.6, 0.6, 0.5)
        Rectangle:
            pos: self.pos
            size: self.size
```

Python File

```python
import kivy.app
import kivy.uix.screenmanager
import kivy.uix.image
import random
import kivy.core.audio
import os
import functools
import kivy.uix.behaviors
import pickle

class TestApp(kivy.app.App):

    def on_start(self):
        music_dir = os.getcwd()+"/music/"
        self.main_bg_music = kivy.core.audio.SoundLoader.load(music_dir+
        "bg_music_piano_flute.wav")
        self.main_bg_music.loop = True
        self.main_bg_music.play()

        next_level_num, congrats_displayed_once = self.read_game_info()
        self.activate_levels(next_level_num, congrats_displayed_once)
```

```python
def read_game_info(self):
    try:
        game_info_file = open("game_info",'rb')
        game_info = pickle.load(game_info_file)
        return game_info[0]['lastlvl'], game_info[0]['congrats_
        displayed_once']
        game_info_file.close()
    except:
        print("CoinTex FileNotFoundError: Game info file is not found.
        Game starts from level 1.")
        return 1, False

def activate_levels(self, next_level_num, congrats_displayed_once):
    num_levels = len(self.root.screens[0].ids['lvls_imagebuttons'].
    children)

    levels_imagebuttons = self.root.screens[0].ids['lvls_
    imagebuttons'].children
    for i in range(num_levels-next_level_num, num_levels):
        levels_imagebuttons[i].disabled = False
        levels_imagebuttons[i].color = [1,1,1,1]

    for i in range(0, num_levels-next_level_num):
        levels_imagebuttons[i].disabled = True
        levels_imagebuttons[i].color = [1,1,1,0.5]

    if next_level_num == (num_levels+1) and congrats_displayed_once ==
    False:
        self.root.current = "alllevelscompleted"

def screen_on_pre_leave(self, screen_num):
    curr_screen = self.root.screens[screen_num]
    for i in range(curr_screen.num_monsters): curr_screen.
    ids['monster'+str(i+1)+'_image_lvl'+str(screen_num)].pos_hint =
    {'x': 0.8, 'y': 0.8}
    curr_screen.ids['character_image_lvl'+str(screen_num)].pos_hint =
    {'x': 0.0, 'y': 0.0}
```

```
        next_level_num, congrats_displayed_once = self.read_game_info()
        self.activate_levels(next_level_num, congrats_displayed_once)

    def screen_on_pre_enter(self, screen_num):
        curr_screen = self.root.screens[screen_num]
        curr_screen.character_killed = False
        curr_screen.num_coins_collected = 0
        curr_screen.ids['character_image_lvl'+str(screen_num)].im_num =
        curr_screen.ids['character_image_lvl'+str(screen_num)].start_im_num
        for i in range(curr_screen.num_monsters): curr_screen.
        ids['monster'+str(i+1)+'_image_lvl'+str(screen_num)].im_num =
        curr_screen.ids['monster'+str(i+1)+'_image_lvl'+str(screen_num)].
        start_im_num
        curr_screen.ids['num_coins_collected_lvl'+str(screen_num)].text =
        "Coins 0/"+str(curr_screen.num_coins)
        curr_screen.ids['level_number_lvl'+str(screen_num)].text = "Level
        "+str(screen_num)

        curr_screen.num_collisions_hit = 0
        remaining_life_percent_lvl_widget = curr_screen.ids['remaining_
        life_percent_lvl'+str(screen_num)]
        remaining_life_percent_lvl_widget.size_hint = (remaining_life_
        percent_lvl_widget.remaining_life_size_hint_x, remaining_life_
        percent_lvl_widget.size_hint[1])

        for i in range(curr_screen.num_fires): curr_screen.
        ids['fire'+str(i+1)+'_lvl'+str(screen_num)].pos_hint = {'x': 1.1,
        'y': 1.1}

        for key, coin in curr_screen.coins_ids.items():
            curr_screen.ids['layout_lvl'+str(screen_num)].remove_
            widget(coin)
        curr_screen.coins_ids = {}

        coin_width = 0.05
        coin_height = 0.05

        curr_screen = self.root.screens[screen_num]
```

```
        section_width = 1.0/curr_screen.num_coins
        for k in range(curr_screen.num_coins):
            x = random.uniform(section_width*k, section_width*(k+1)-coin_
            width)
            y = random.uniform(0, 1-coin_height)
            coin = kivy.uix.image.Image(source="coin.png", size_hint=(coin_
            width, coin_height), pos_hint={'x': x, 'y': y}, allow_
            stretch=True)
            curr_screen.ids['layout_lvl'+str(screen_num)].add_widget(coin,
            index=-1)
            curr_screen.coins_ids['coin'+str(k)] = coin

    def screen_on_enter(self, screen_num):
        music_dir = os.getcwd()+"/music/"
        self.bg_music = kivy.core.audio.SoundLoader.load(music_dir+"bg_
        music_piano.wav")
        self.bg_music.loop = True

        self.coin_sound = kivy.core.audio.SoundLoader.load(music_dir+"coin.
        wav")
        self.level_completed_sound = kivy.core.audio.SoundLoader.
        load(music_dir+"level_completed_flaute.wav")
        self.char_death_sound = kivy.core.audio.SoundLoader.load(music_
        dir+"char_death_flaute.wav")

        self.bg_music.play()

        curr_screen = self.root.screens[screen_num]
        for i in range(curr_screen.num_monsters):
            monster_image = curr_screen.ids['monster'+str(i+1)+'_image_
            lvl'+str(screen_num)]
            new_pos = (random.uniform(0.0, 1 - monster_image.size_
            hint[0]/4), random.uniform(0.0, 1 - monster_image.size_
            hint[1]/4))
            self.start_monst_animation(monster_image=monster_image, new_
            pos=new_pos, anim_duration=random.uniform(monster_image.monst_
            anim_duration_low, monster_image.monst_anim_duration_high))
```

```python
    for i in range(curr_screen.num_fires):
        fire_widget = curr_screen.ids['fire'+str(i+1)+'_lvl'+
        str(screen_num)]
        self.start_fire_animation(fire_widget=fire_widget, pos=(0.0,
        0.5), anim_duration=5.0)

def start_monst_animation(self, monster_image, new_pos, anim_duration):
    monst_anim = kivy.animation.Animation(pos_hint={'x': new_pos[0],
    'y': new_pos[1]}, im_num=monster_image.end_im_num,duration=anim_
    duration)
    monst_anim.bind(on_complete=self.monst_animation_completed)
    monst_anim.start(monster_image)

def monst_animation_completed(self, *args):
    monster_image = args[1]
    monster_image.im_num = monster_image.start_im_num

    new_pos = (random.uniform(0.0, 1 - monster_image.size_hint[0]/4),
    random.uniform(0.0, 1 - monster_image.size_hint[1]/4))
    self.start_monst_animation(monster_image=monster_image, new_pos=
    new_pos,anim_duration=random.uniform(monster_image.monst_anim_
    duration_low, monster_image.monst_anim_duration_high))

def monst_pos_hint(self, monster_image):
    screen_num = int(monster_image.parent.parent.name[5:])
    curr_screen = self.root.screens[screen_num]
    character_image = curr_screen.ids['character_image_lvl'+str(screen_
    num)]

    character_center = character_image.center
    monster_center = monster_image.center

    gab_x = character_image.width / 2
    gab_y = character_image.height / 2
    if character_image.collide_widget(monster_image) and abs(character_
    center[0] - monster_center[0]) <= gab_x and abs(character_
    center[1] - monster_center[1]) <= gab_y:
```

```
        curr_screen.num_collisions_hit = curr_screen.num_collisions_hit
        + 1
        life_percent = float(curr_screen.num_collisions_hit)/
        float(curr_screen.num_collisions_level)
#        life_remaining_percent = 100-round(life_percent, 2)*100
#        remaining_life_percent_lvl_widget.text = str(int(life_
        remaining_percent))+"%"
        remaining_life_percent_lvl_widget=curr_screen.ids['remaining_
        life_percent_lvl'+str(screen_num)]
        remaining_life_size_hint_x = remaining_life_percent_lvl_widget.
        remaining_life_size_hint_x
        remaining_life_percent_lvl_widget.size_hint =  (remaining_life_
        size_hint_x-remaining_life_size_hint_x*life_percent, remaining_
        life_percent_lvl_widget.size_hint[1])

        if curr_screen.num_collisions_hit == curr_screen.num_
        collisions_level:
            self.bg_music.stop()
            self.char_death_sound.play()
            curr_screen.character_killed = True

            kivy.animation.Animation.cancel_all(character_image)
            for i in range(curr_screen.num_monsters): kivy.animation.
            Animation.cancel_all(curr_screen.ids['monster'+str(i+1)+'_
            image_lvl'+str(screen_num)])
            for i in range(curr_screen.num_fires): kivy.animation.
            Animation.cancel_all(curr_screen.ids['fire'+str(i+1)+
            '_lvl'+str(screen_num)])

            character_image.im_num = character_image.dead_start_im_num
            char_anim = kivy.animation.Animation(im_num=character_
            image.dead_end_im_num, duration=1.0)
            char_anim.start(character_image)
            kivy.clock.Clock.schedule_once(functools.partial(self.back_
            to_main_screen, curr_screen.parent), 3)
```

```python
def change_monst_im(self, monster_image):
    monster_image.source = str(int(monster_image.im_num)) + ".png"

def touch_down_handler(self, screen_num, args):
    curr_screen = self.root.screens[screen_num]
    if curr_screen.character_killed == False:
        self.start_char_animation(screen_num, args[1].spos)

def start_char_animation(self, screen_num, touch_pos):
    curr_screen = self.root.screens[screen_num]
    character_image = curr_screen.ids['character_image_lvl'+str(screen_
num)]
    character_image.im_num = character_image.start_im_num
    char_anim = kivy.animation.Animation(pos_hint={'x': touch_pos[0] -
    character_image.size_hint[0] / 2,'y': touch_pos[1] - character_
    image.size_hint[1] / 2}, im_num=character_image.end_im_num,
    duration=curr_screen.char_anim_duration)
    char_anim.bind(on_complete=self.char_animation_completed)
    char_anim.start(character_image)

def char_animation_completed(self, *args):
    character_image = args[1]
    character_image.im_num = character_image.start_im_num

def char_pos_hint(self, character_image):
    screen_num = int(character_image.parent.parent.name[5:])
    character_center = character_image.center

    gab_x = character_image.width / 3
    gab_y = character_image.height / 3
    coins_to_delete = []
    curr_screen = self.root.screens[screen_num]

    for coin_key, curr_coin in curr_screen.coins_ids.items():
        curr_coin_center = curr_coin.center
        if character_image.collide_widget(curr_coin) and abs(character_
        center[0] - curr_coin_center[0]) <= gab_x and abs(character_
        center[1] - curr_coin_center[1]) <= gab_y:
```

```
self.coin_sound.play()
coins_to_delete.append(coin_key)
curr_screen.ids['layout_lvl'+str(screen_num)].remove_
widget(curr_coin)
curr_screen.num_coins_collected = curr_screen.num_coins_
collected + 1
curr_screen.ids['num_coins_collected_lvl'+str(screen_
num)].text = "Coins "+str(curr_screen.num_coins_
collected)+"/"+str(curr_screen.num_coins)
if curr_screen.num_coins_collected == curr_screen.num_
coins:
    self.bg_music.stop()
    self.level_completed_sound.play()
    kivy.clock.Clock.schedule_once(functools.partial(self.
    back_to_main_screen, curr_screen.parent), 3)
    for i in range(curr_screen.num_monsters): kivy.
    animation.Animation.cancel_all(curr_screen.
    ids['monster'+str(i+1)+'_image_lvl'+str(screen_num)])
    for i in range(curr_screen.num_fires): kivy.animation.
    Animation.cancel_all(curr_screen.ids['fire'+str(i+1)+
    '_lvl'+str(screen_num)])

    next_level_num, congrats_displayed_once = self.read_
    game_info()
    if (screen_num+1) > next_level_num:
        game_info_file = open("game_info",'wb')
        pickle.dump([{'lastlvl':screen_num+1, "congrats_
        displayed_once": False}], game_info_file)
        game_info_file.close()
    else:
        game_info_file = open("game_info",'wb')
        pickle.dump([{'lastlvl':next_level_num, "congrats_
        displayed_once": True}], game_info_file)
        game_info_file.close()
```

```python
        if len(coins_to_delete) > 0:
            for coin_key in coins_to_delete:
                del curr_screen.coins_ids[coin_key]

    def change_char_im(self, character_image):
        character_image.source = str(int(character_image.im_num)) + ".png"

    def start_fire_animation(self, fire_widget, pos, anim_duration):
        fire_anim = kivy.animation.Animation(pos_hint=fire_widget.fire_
        start_pos_hint, duration=fire_widget.fire_anim_duration)+kivy.
        animation.Animation(pos_hint=fire_widget.fire_end_pos_hint,
        duration=fire_widget.fire_anim_duration)
        fire_anim.repeat = True
        fire_anim.start(fire_widget)

    def fire_pos_hint(self, fire_widget):
        screen_num = int(fire_widget.parent.parent.name[5:])
        curr_screen = self.root.screens[screen_num]
        character_image = curr_screen.ids['character_image_lvl'+str(screen_
        num)]

        character_center = character_image.center
        fire_center = fire_widget.center

        gab_x = character_image.width / 3
        gab_y = character_image.height / 3
        if character_image.collide_widget(fire_widget) and abs(character_
        center[0] - fire_center[0]) <= gab_x and abs(character_center[1] -
        fire_center[1]) <= gab_y:
            curr_screen.num_collisions_hit = curr_screen.num_collisions_hit
            + 1
            life_percent = float(curr_screen.num_collisions_hit)/
            float(curr_screen.num_collisions_level)

            remaining_life_percent_lvl_widget = curr_screen.ids['remaining_
            life_percent_lvl'+str(screen_num)]
#           life_remaining_percent = 100-round(life_percent, 2)*100
#           remaining_life_percent_lvl_widget.text = str(int(life_
            remaining_percent))+"%"
```

```
            remaining_life_size_hint_x = remaining_life_percent_lvl_widget.
            remaining_life_size_hint_x
            remaining_life_percent_lvl_widget.size_hint = (remaining_life_
            size_hint_x-remaining_life_size_hint_x*life_percent, remaining_
            life_percent_lvl_widget.size_hint[1])

            if curr_screen.num_collisions_hit == curr_screen.num_
            collisions_level:
                self.bg_music.stop()
                self.char_death_sound.play()
                curr_screen.character_killed = True

                kivy.animation.Animation.cancel_all(character_image)
                for i in range(curr_screen.num_monsters): kivy.animation.
                Animation.cancel_all(curr_screen.ids['monster'+str(i+1)+
                '_image_lvl'+str(screen_num)])
                for i in range(curr_screen.num_fires): kivy.animation.
                Animation.cancel_all(curr_screen.ids['fire'+str(i+1)+
                '_lvl'+str(screen_num)])

                character_image.im_num = character_image.dead_start_im_num
                char_anim = kivy.animation.Animation(im_num=character_
                image.dead_end_im_num, duration=1.0)
                char_anim.start(character_image)
                kivy.clock.Clock.schedule_once(functools.partial(self.back_
                to_main_screen, curr_screen.parent), 3)

    def back_to_main_screen(self, screenManager, *args):
        screenManager.current = "main"

    def main_screen_on_enter(self):
        self.main_bg_music.play()

    def main_screen_on_leave(self):
        self.main_bg_music.stop()

class ImageButton(kivy.uix.behaviors.ButtonBehavior, kivy.uix.image.Image):
    pass
```

```python
class MainScreen(kivy.uix.screenmanager.Screen):
    pass

class AboutUs(kivy.uix.screenmanager.Screen):
    pass

class AllLevelsCompleted(kivy.uix.screenmanager.Screen):
    pass

class Level1(kivy.uix.screenmanager.Screen):
    character_killed = False
    num_coins = 5
    num_coins_collected = 0
    coins_ids = {}
    char_anim_duration = 1.0
    num_monsters = 1
    num_fires = 0
    num_collisions_hit = 0
    num_collisions_level = 20

class Level2(kivy.uix.screenmanager.Screen):
    character_killed = False
    num_coins = 8
    num_coins_collected = 0
    coins_ids = {}
    char_anim_duration = 1.1
    num_monsters = 1
    num_fires = 0
    num_collisions_hit = 0
    num_collisions_level = 30

class Level3(kivy.uix.screenmanager.Screen):
    character_killed = False
    num_coins = 12
    num_coins_collected = 0
    coins_ids = {}
    char_anim_duration = 1.2
    num_monsters = 1
```

```python
    num_fires = 0
    num_collisions_hit = 0
    num_collisions_level = 30

class Level4(kivy.uix.screenmanager.Screen):
    character_killed = False
    num_coins = 10
    num_coins_collected = 0
    coins_ids = {}
    char_anim_duration = 1.2
    num_monsters = 1
    num_fires = 1
    num_collisions_hit = 0
    num_collisions_level = 20

class Level5(kivy.uix.screenmanager.Screen):
    character_killed = False
    num_coins = 15
    num_coins_collected = 0
    coins_ids = {}
    char_anim_duration = 1.3
    num_monsters = 1
    num_fires = 2
    num_collisions_hit = 0
    num_collisions_level = 20

class Level6(kivy.uix.screenmanager.Screen):
    character_killed = False
    num_coins = 12
    num_coins_collected = 0
    coins_ids = {}
    char_anim_duration = 1.3
    num_monsters = 1
    num_fires = 3
    num_collisions_hit = 0
    num_collisions_level = 20
```

```python
class Level7(kivy.uix.screenmanager.Screen):
    character_killed = False
    num_coins = 10
    num_coins_collected = 0
    coins_ids = {}
    char_anim_duration = 1.4
    num_monsters = 3
    num_fires = 0
    num_collisions_hit = 0
    num_collisions_level = 25

class Level8(kivy.uix.screenmanager.Screen):
    character_killed = False
    num_coins = 15
    num_coins_collected = 0
    coins_ids = {}
    char_anim_duration = 1.4
    num_monsters = 2
    num_fires = 0
    num_collisions_hit = 0
    num_collisions_level = 25

class Level9(kivy.uix.screenmanager.Screen):
    character_killed = False
    num_coins = 12
    num_coins_collected = 0
    coins_ids = {}
    char_anim_duration = 1.5
    num_monsters = 2
    num_fires = 0
    num_collisions_hit = 0
    num_collisions_level = 25

class Level10(kivy.uix.screenmanager.Screen):
    character_killed = False
    num_coins = 14
    num_coins_collected = 0
```

```
    coins_ids = {}
    char_anim_duration = 1.5
    num_monsters = 3
    num_fires = 0
    num_collisions_hit = 0
    num_collisions_level = 30

class Level11(kivy.uix.screenmanager.Screen):
    character_killed = False
    num_coins = 15
    num_coins_collected = 0
    coins_ids = {}
    char_anim_duration = 1.6
    num_monsters = 2
    num_fires = 1
    num_collisions_hit = 0
    num_collisions_level = 30

class Level12(kivy.uix.screenmanager.Screen):
    character_killed = False
    num_coins = 12
    num_coins_collected = 0
    coins_ids = {}
    char_anim_duration = 1.6
    num_monsters = 2
    num_fires = 1
    num_collisions_hit = 0
    num_collisions_level = 30

class Level13(kivy.uix.screenmanager.Screen):
    character_killed = False
    num_coins = 10
    num_coins_collected = 0
    coins_ids = {}
    char_anim_duration = 1.7
    num_monsters = 2
    num_fires = 2
```

```
    num_collisions_hit = 0
    num_collisions_level = 20

class Level14(kivy.uix.screenmanager.Screen):
    character_killed = False
    num_coins = 15
    num_coins_collected = 0
    coins_ids = {}
    char_anim_duration = 1.7
    num_monsters = 0
    num_fires = 6
    num_collisions_hit = 0
    num_collisions_level = 30

class Level15(kivy.uix.screenmanager.Screen):
    character_killed = False
    num_coins = 16
    num_coins_collected = 0
    coins_ids = {}
    char_anim_duration = 1.8
    num_monsters = 2
    num_fires = 3
    num_collisions_hit = 0
    num_collisions_level = 30

class Level16(kivy.uix.screenmanager.Screen):
    character_killed = False
    num_coins = 15
    num_coins_collected = 0
    coins_ids = {}
    char_anim_duration = 1.8
    num_monsters = 3
    num_fires = 2
    num_collisions_hit = 0
    num_collisions_level = 35
```

```python
class Level17(kivy.uix.screenmanager.Screen):
    character_killed = False
    num_coins = 10
    num_coins_collected = 0
    coins_ids = {}
    char_anim_duration = 1.3
    num_monsters = 0
    num_fires = 4
    num_collisions_hit = 0
    num_collisions_level = 30

class Level18(kivy.uix.screenmanager.Screen):
    character_killed = False
    num_coins = 15
    num_coins_collected = 0
    coins_ids = {}
    char_anim_duration = 1.5
    num_monsters = 3
    num_fires = 4
    num_collisions_hit = 0
    num_collisions_level = 30

class Level19(kivy.uix.screenmanager.Screen):
    character_killed = False
    num_coins = 12
    num_coins_collected = 0
    coins_ids = {}
    char_anim_duration = 1.2
    num_monsters = 0
    num_fires = 6
    num_collisions_hit = 0
    num_collisions_level = 30

class Level20(kivy.uix.screenmanager.Screen):
    character_killed = False
    num_coins = 15
    num_coins_collected = 0
```

```python
    coins_ids = {}
    char_anim_duration = 1.1
    num_monsters = 0
    num_fires = 8
    num_collisions_hit = 0
    num_collisions_level = 30

class Level20(kivy.uix.screenmanager.Screen):
    character_killed = False
    num_coins = 20
    num_coins_collected = 0
    coins_ids = {}
    char_anim_duration = 1.1
    num_monsters = 2
    num_fires = 4
    num_collisions_hit = 0
    num_collisions_level = 30

class Level21(kivy.uix.screenmanager.Screen):
    character_killed = False
    num_coins = 18
    num_coins_collected = 0
    coins_ids = {}
    char_anim_duration = 1.3
    num_monsters = 2
    num_fires = 4
    num_collisions_hit = 0
    num_collisions_level = 30

class Level22(kivy.uix.screenmanager.Screen):
    character_killed = False
    num_coins = 20
    num_coins_collected = 0
    coins_ids = {}
    char_anim_duration = 1.3
    num_monsters = 2
    num_fires = 4
```

```
    num_collisions_hit = 0
    num_collisions_level = 30

class Level23(kivy.uix.screenmanager.Screen):
    character_killed = False
    num_coins = 25
    num_coins_collected = 0
    coins_ids = {}
    char_anim_duration = 1.1
    num_monsters = 2
    num_fires = 2
    num_collisions_hit = 0
    num_collisions_level = 30

class Level24(kivy.uix.screenmanager.Screen):
    character_killed = False
    num_coins = 20
    num_coins_collected = 0
    coins_ids = {}
    char_anim_duration = 1.1
    num_monsters = 3
    num_fires = 2
    num_collisions_hit = 0
    num_collisions_level = 30

app = TestApp(title="CoinTex")
app.run()
```

Index

A

activate_levels() method, 290

<activity> element, 313

addView() method, 343

add_widget function, 23, 178

Android application

    assign ID, 65

    capture and upload image, 64, 65

    client-side application, 96

    live camera, 93, 95

    server-side application, 96

Android Camera

    Buildozer, 41

    image capture, 42

android:installLocation, 310

android.ndk_path property, 10

android.permissions property, 9, 298, 310

Android SDK/NDK, 13–16

android.sdk_path property, 10

android:versionCode property, 309

Animation

    cancel_all() function, 138–140

    create application, image widget, 131

    join, 135, 137

    separating multiple size, 133, 134

    source property of image widget

        adding im_num argument, 141

        adding im_num event, 142

        adding im_num property, 141

        change_char_im() function, 143–145

        on_num event, 143

        steps, 141

    start_char_animation() function, 132

<application> element, 311–313

autoclass() function, 299, 302

B

b64encode() function, 86

back_to_main_screen() function, 234

bind() function, 25

BoxLayout root widget, 40, 103

Kivy BoxLayout container, 56

build() function, 4, 21, 23, 34, 40, 108, 178, 181, 203, 204, 219

Buildozer

    Android

        AndroidManifest.xml file, 313

        screenOrientation, 313

        versionCode property, 309

    AndroidManifest.templ.xml file, 11, 12

    AndroidManifest.xml file, 309–311

    build.gradle, 309

    buildozer.spec file, 6, 7, 10, 55

    build.tmpl.gradle file, 10, 11

    installation, 5

    mLayout variable, 318

    NewApp, 6

    PythonActivity, 314

415

© Ahmed Fawzy Mohamed Gad 2019

A. F. M. Gad, *Building Android Apps in Python Using Kivy with Android Studio*,
https://doi.org/10.1007/978-1-4842-5031-0

Printed in the United States
By Bookmasters

Printed in the United States
By Bookmasters